Visit **www.parentsforafuture.org** to keep up with citizen action related to this book.

@Parents4aFuture #ParentsForaFuture

More praise for Rupert Read and 'Parents For A Future'

'Rupert Read is more than just a brilliant thinker. He is a determined and responsible campaigner for a better country and a better world. I've known him since the legendary Newbury bypass protests over a generation ago now. He has never abandoned his political commitments, or his commitment to humanity and the rest of the living world.'
– George Monbiot

'In the absence of grandmothers and grand uncles, modernity is depriving us of intergenerational precautionary wisdom. Thankfully, Rupert Read is stepping in to fill in that role, with clarity... and courage.'
– Nassim Nicholas Taleb

'There are not too many people on this planet, even among those most alarmed about global warming, who are actually living as though they believe that the climate is in genuine crisis and that every promise and hope we have ever extended for future human flourishing is in doubt as a result. Rupert Read is one of them. This book is his inspired, incisive, mobilizing endeavour to recruit more of us to the urgent cause.'
– David Wallace-Wells

'*Parents for A Future* is a compelling, confronting and ultimately uplifting book by one of the most important radical thinkers of our time. With a scintillating combination of philosophical rigour, political urgency and moral vision, Rupert Read reveals why caring for our loved ones is a stepping stone to caring about a long-term future for both people and planet. A clear-sighted and necessary handbook for reinventing our ethical and political landscape at a time of global crisis.'
– Roman Krznaric, author of *The Good Ancestor*

'An emergency is *now*, but for most people the climate crisis is still a worry about the future, so no wonder we aren't tackling it urgently enough. Rupert Read, philosopher and activist, shows how ordinary love for our children in the present can motivate the radical changes which might yet give us a chance of avoiding catastrophe. This is philosophy for once doing its proper job, helping us all think straight about things that matter vitally.'
– **John Foster, author of *After Sustainability***

'Rupert Read is an excellent philosopher; educated at top universities, he remains outstanding among his peers. In part this is because he is a rare philosopher: he's someone who is actually determined to change the world (not merely interpret it). I admire the way he has put himself on the line for the sake of our common future: up to the point even of being arrested for the cause.'
– **Anne Jaap Jacobson, Professor Emeritus of Philosophy and Engineering, UH**

'Rupert Read is one of the few people of my generation who brings an emotional intensity to the page that recalls an older existentialist attitude toward philosophy. Read is a living reminder that ideas matter.'
– **Steve Fuller, author of *Humanity 2.0***

'If you believe that humanity is fundamentally about caring for ourselves and others, then this book is simply a 'must read'. It will walk you through a series of logical steps that will convince you to act to prevent the death of our planet and with it, the death of the human future. Please read this book as if your life and the life of everyone you care about depends on it.'
– **Etienne Stott, rower, Olympic gold medallist London 2012**

'This is an amazing, deeply thoughtful book from one of the world's leading climate activists that shines an uncompromising spotlight on our descent into climate chaos but shows how we can extend the powerful love and duty of care we feel for our families outwards to nurture the precious planet we call home.'
– **Claire Perry O'Neill, Former UK Climate, Clean Growth and Energy Minister & COP26 President-Designate; MD at the World Business Council for Sustainable Development**

'Confronting head on the uncomfortable reality that most people are still not listening to the science of climate change — with all its dreadful implications for the future of humankind — Rupert Read has chosen a different path in *Parents for a Future*, appealing to people's emotions and their deeper sense of themselves. This is a moving and powerful response to Greta Thunberg's injunction to all of us that we should 'Act as if you loved your children above all else'.'
– **Jonathon Porritt, author of *Hope in Hell***

'With his unique brand of fierce and compassionate logic, Rupert Read compels us to face the truth: loving our children means doing everything in our power to protect the ecological and climate systems we are part of and utterly depend on. And that means going well beyond individual lifestyle change to a profound and urgently needed reworking of our political and economic systems.'
– **Kate Rawles, author, environmentalist and outdoor philosopher**

'A powerful, passionate call for intergenerational solidarity. Read it; weep; then act.'
– **Stephen M. Gardiner, author of *A Perfect Moral Storm: The Ethical Tragedy of Climate Change***

Parents For A Future

Parents For A Future

How Loving Our Children Can Prevent
Climate Collapse

UEA
PUBLISHING
PROJECT

'We do not err because truth is difficult to see. It is visible at a glance. We err because it is more comfortable.'
– Aleksandr Solzhenitsyn

Foreword

'What did you do to help stop climate change?'

As a parent, this is a question you are likely to be called on to answer by your children and grandchildren in the coming decades.

How will you respond?

As a parent myself, of a cheeky six-year-old boy, I often wonder about this question — as I know many other parents do. So far, we've collectively failed when it comes to even slowing down, let alone reversing, catastrophic climate change.

Confronting this reality feels overwhelming. However, as Rupert Read points out, it's part of our human conditioning — to intellectually understand a future crisis, but not to feel the urgency to act now.

As Mark Lynas says in the film *The Age of Stupid*, humans did not evolve to be worried about issues years in the future.

The only place where Rupert observes an exception — and as a parent I can attest to this — is with our children. I am willing to sacrifice, take bold action and move mountains to give my son the best future possible. I love him so much. I want him to avoid all the pain I experienced growing up.

I'm sure as a parent (or guardian or fosterer, or aunt or uncle) you know exactly what I'm talking about.

When we think about protecting our children, changes that would normally seem too drastic for us (such as not eating meat, not flying and getting an electric car) seem possible. Taking the next step to persuade friends and family to do the same feels worthwhile.

Rupert directly addresses our inner sceptic that pipes up when we consider changing our diets and purchasing habits... 'Surely if the situation were really as alarming as Extinction Rebellion and Greta Thunberg make out, then surely governments would act? I'm sure they'll sort this out.'

I write this in the second UK lockdown of the Covid-19 pandemic. Despite governments knowing for decades that pandemics are a big risk in our ever globalised world, it is devastatingly apparent that we in the UK were unprepared.

Many of us are looking over to Germany, where they have twenty million more people but (at the time of writing) only a sixth of the number of deaths. While they were preparing a Covid-19 test in January 2020, the UK Government was arguing about whether 'Big Ben should Bong for Brexit'.

This is a perfect example of both how distracted and short term our thinking is. And, how our children's and grandchildren's future could be decided on how many populist governments are elected around the world.

It's easy to be overwhelmed, fearful and unsure of what to do next. But as parents we are used to handling these emotions — especially when our children are very young. We stepped up for them then, and they need us to rise up and take action now.

I suspect you picked up this book because you are looking for actions. One that I say to anyone wanting to do something right away is to go to Ecologi.com — for the same price as a Netflix subscription you and your family can become climate positive through tree planting and carbon credits.

But while this is a great place to start, the journey Rupert is about to take you on requires courage and ambition to think bigger and bolder.

You will likely find some points in this book challenging. However, my hope is that many of these ideas actually inspire you, and help you consider different ways we could find solutions to this crisis.

Most of all, I hope that you find your own path to answer that future question: what did you do to stop climate change?

For me, as the co-founder of a successful and growing business, I've committed us to planting over one million trees by 2025 to draw down C02 way beyond my business' collective carbon footprint. I've made this commitment to personally plant a million trees as well. These will cover hundreds of people who can't afford to do this themselves. My goal is to persuade over one thousand individuals and businesses to take the one million tree pledge as well.

I'm committed to doing this for all of us, but, most importantly, I'm compelled to do this for my six-year-old son, Zac.

Marcus Hemsley
Co-founder of Fountain Partnership

#RiseUpDrawDown

1

S.O.S.: Save Our Species

'What has to be overcome are not difficulties of the intellect but of the will.'
– Ludwig Wittgenstein[1]

On trying

Reader, I want to invite you into an essay proposing a new way to address the great issue of our time: how — even now, at the twelfth hour — we might turn ourselves away from our current path of self-destruction.

Even as I mount this attempt, I am daunted by the scale of the task. Perhaps it daunts you too. And I am unnerved by the seeming inadequacy of the means at our disposal.

How can one even begin to write an essay on something as big, as incomprehensible, as the end of our world — let alone propose a way of preventing it? For one thing, isn't the essay, as a literary form, too assumptive of something like the status quo, too redolent of a parlour game? Too 'clever', too self-indulgent and brief? Too, well... slight?

And how can a mere essay possibly succeed where far greater forces have failed? Haven't we — and by 'we' I mean everyone: scientists, ecologists, politicians, intellectuals, not to mention many active citizens — already attempted to present the evidence, marshal the arguments and press for the policies? Haven't we — and by 'we' I mean now our society itself — already shown ourselves terminally incapable of the radical action needed? After all, despite everything we know, despite everything science has told us, despite even the opportunity that was afforded us by the coronavirus crisis for a radical reset, we remain firmly on course for burn-up, for an ecologically induced societal collapse.[2]

A strange phenomenon: facts, no matter how unequivocal, don't seem motivate us.

If facts haven't been able to shift the needle, change our ways of thinking and galvanise us into meaningful action, what about fiction? Here too, where one might have hoped for a deeper engagement, we tend to find a dearth of it. Take TV. Television remains the most influential mass medium of our time. It is incredible — yet true — that there has been no significant fictional television series addressing front and centre the climate and ecological emergency.[3] (The recent David Attenborough documentaries, *Climate Change: The Facts* (2019) and *A Life On Our Planet* (2020), have made some real impact. They put fiction/art to shame.) Nothing by way of story, on the main media in the world, focussed on the biggest issue of all. More generally, the arts — a space where, we might think, the imagination can be harnessed to make the impossible seem possible — have so far largely failed to face this issue adequately, failed to give us tools with which to think through our predicament. Possible exceptions to that judgement might include *The Road* (2009), *Melancholia* (2011) and *Avatar* (2009), with an honourable mention too for *Wall-E* (2008).[4] These three fine works of film (with *The Road* having started life as a magnificent novel) do provide ways of starting to approach this matter that matters most of all. But, for all their inspirational qualities, they are unlikely to strike anyone as laying out a realistic path for how we are to save ourselves. (*The Road* and *Melancholia* find a fragment of redemption in the face of utter destruction, while *Avatar* ends in marvellously unrealistic fashion by positing a Gaia-like 'god' awakening to become the saving power.)

Amitav Ghosh, in *The Great Derangement* (2016),[5] has shown beautifully how and why the novel as we know it is too 'realistic' a medium to tackle this topic. In the nineteenth-century heyday of the realist novel, someone like Charles Dickens could use fiction as a vehicle for tackling the pressing issues of the day. In our time, as Ghosh suggests, the inconceivably vast, sublimely mind-blowing challenge of human-induced climate breakdown exceeds the bounds of realist description. It departs from what we thought we knew our world to be like and strikes off into a terrifying unknown. It exceeds reality as we know it.

As our Anthropocenic weather[6] gradually goes psychotic, the very categories of '*natural*' disaster or 'act of god' — categories that we've relied upon for so long — become unavailable to us; because it is humanity's effects that now pattern the whole. There is nothing supernatural about the utterly unnatural climate we are entering.

Ghosh's book shows how the uncanniness of global weirding is incompatible with the norms of the novel.[7] (It might just be parsable within those of an epic.) Ghosh's founding example is his own experience of the first ever tornado to hit Delhi: a tornado so strange and novel that he has never found a way to integrate the experience of it into his novels.

The truth — the reality — is that the phenomena of man-made deadly climate change really are stranger than fiction. If we continue on our current path, the terrible future that awaits us will go on striking most readers as 'unrealistic', 'unbelievable' — fodder (it would seem) for 'escapist' blockbusters, but not literary realism, let alone actual reality. My point: unless we can find a way of envisaging that which is coming our way, unless we find a way of believing in the reality of climate breakdown, then we will not succeed in averting it.

It's an awful paradox: our inability to believe in this overwhelmingly likely future assures we remain on course for it. As things stand, we simply can't bring ourselves to credit that which science tells us is true or likely. It's too vast, too deeply strange, or just too awful to look in the eye. And yet, there will be no escape from it, *unless* we look it in the eye — face it and change everything to swerve our future from it.

Of course, we have recently had a shot across the bow, one which even yet might just occasion the massive realignment we need. The terrible, broadly foreseeable outbreak[8] of Covid-19 is itself a product of the ecological crisis: of habitat destruction for the sake of economic growth,[9] of the maltreatment of animals,[10] of economic globalisation.[11] It has made the unimaginable much more imaginable than it was, because, in one key respect, we don't even have to imagine it: it's here. We are, in this long moment, undergoing an experience of planetary emergency, of lived vulnerability, of potential mortality: our own, our parents', our grandparents'; just conceivably, our society's. For many of us, especially in the Global North, it's the first time this has happened in our lifetimes.

Furthermore, we have seen many governments act with incredible speed and boldness, spending money like water. (Never again will the lack of public money be a tenable reason for states not taking major protective action in the face of an emergency.) We experienced a re-attunement to the value of nature, as we heard more birdsong under lockdown. Many of us have realised that it isn't necessary to commute after all. And here where I am writing, in the UK, we have come to re-appreciate the value of care, the marvel of the NHS, the starkness of the economic divisions fracturing our society, the centrality to us of love. With great vulnerability comes great responsibility — and great power.

Within the horror of coronavirus, then, comes a concealed gift. If we can transfer this sense of vulnerability to the larger emergency[12] — the ecological crisis that is parent to the pandemic — then we will have taken a huge step forward. After the December 2019 general election in the UK, I felt very discouraged. With the big victory of a melting block ice as Prime Minister, we had likely missed our last shot at the political system hitting the targets for transformational change that might have headed off eco-driven societal collapse. Covid-19, and the years of reset it will require, may have given us one last chance to do what this book asks of us. These opportunities don't come often; the last was the financial crisis of 2008 and we allowed it to pass by. We cannot allow that to happen again. And yet it is happening: we have returned to an alarming extent to business as usual, bailing out the polluters, resuming car travel and even air travel. We've lost much of the opportunity afforded by the Covid crisis. If we let it go by, then we will have lost our very last chance at transforming rather than destroying our civilisation.[13]

So we can and must but try... And in this way I've started to make my case. Despite the litany of failure, of insufficiency of one kind or another, individual or collective, this book sets out to make the argument finally stick, the proposal hold, the solution a thing you can believe in.

In the verb I was drawn to above — *try* — we might find a justification for my choice of form. After all, an essay is an attempt, a try — a having a go at something. And we have to try, even when — in fact, precisely *because* — all those other methods have failed. For the only thing we know for sure is that if we stop trying to imagine an answer[14] — if we give up essaying something truly bold, something ingenious and ingenuous enough to truly tackle the long emergency — then we are certain to face a catastrophic civilisational decline. If those of us seeking a way to stop the coming climate cataclysm give up trying, then it is certain that that cataclysm will come. For it is heading our way, as surely as Hurricane Katrina headed for New Orleans. And so we have this thing I'm sharing with you: *my* try; my essay.

Of course, even trying as hard as we can may not be enough. The transformations our civilisation requires will be astonishingly hard to achieve and the obstacles we face are formidable: from chronic short-termism to entrenched fossil interests, from a deadly time-lag to waning attention spans. The corona crisis doesn't only help us, it also gets in the way: look at the difficulties public transport is now in; look at the way the digital behemoths and the surveillance state have been strengthened; look at the way in which climate-bandwidth is now being taken up by corona-preoccupation.[15]

We are on course to fail, and this supertanker will take the almightiest turning. Life is so very precious, and we see that all the more clearly when we know it may be short.

So, enjoy it while it lasts. And that means: do the right thing, regardless of outcomes. Don't 'attach' to the outcomes you hope for. In fact, give up ordinary hope. It's too late for ordinary optimism, for waiting for something or someone to come along, just as it's too late for pessimism to be anything other than an evasion. All that's left now is a deeper realism — and action.

If (and only if) we start really trying, then hope will sprout everywhere. I essay here a way that we might finally put ourselves in a position to see adequately what is to be done, and why it must be done.

The idea that we must safeguard the environment for future generations is a commonplace. But as yet it has almost completely lacked any felt binding power upon almost any of us. I essay here the bringing of what we already sense into a place of deeper knowing. So that we might finally put ourselves in a position where we can see what is to be done and why it must be done, and so that our actions — in all our life roles, and despite all of life's complexity — can flow from this place of knowing.

Parenting the future

What is the nub of my case for how we can tackle this, the great issue of our time? We need to become parents of the future. How do we do that? By taking the metaphor literally. By understanding that only if we take completely seriously what it *really* means to be good parents to our children, will it *really* be enough to take care of the whole future.

What I'm trying to do, by marshalling this particular metaphor, is find a way of helping us face something that we don't want to face: the destruction we are wreaking on ourselves and, with greater finality, on our children. We don't want to face it, because to do so will mean we have to grieve deeply — and to change our lives profoundly. We need to mourn all the nature that is lost, mourn our lost innocence, perhaps mourn the loss of the life and material 'progress' we had hoped for; and even mourn pre-emptively for the horrible losses to come. This mourning-work clears the ground upon which we can and must then stand, unwilling to allow our common future to be smashed. Present, with determination, to undertake the great transformation.[16]

By focussing our minds on the young vulnerable generation that we lavish our love upon, I'm trying to find a way into the problem that will appeal not only to anyone who's ever had children, but to anyone who understands the duty of care that parenthood entails. Our sense of this duty, I argue, is so strong and sweet that it can, as it were, move the Earth. It is a way of feeling, of waking up, that can move us and, in moving us, will move our children and their children out of the firing line, out of harm's way.

If we allow ourselves so to be moved, then finally we have a place to stand, a place to make our stand. Within the Earth, not 'outside' it. Within reality, not evading it any longer.

This calls us to face climate reality. To face the truth. As people have woken up to what Trump and the Brexit referendum have done to our politics and our civic life over recent years, there is a growing pushback against the inanities and insanities of 'post-truth',[17] a pushback accelerated by the realisation, with the coronavirus, that we haven't had enough (of) experts. Countries whose governments refused to admit the reality of the coronavirus outbreak — the USA, Brazil, Sweden, and yes, the UK — have seen their populations and economies needlessly ravaged. There is a growing awareness of the need to defend facts and science if we want to remain in the gene pool. This is a hopeful sign. If we can learn again to track the truth and to tell it, then we'll have passed first base.

But once again, facts (as opposed to the alt-variety) are not going to be enough. The task I am engaging in here requires something much more than responsiveness to evidence. The evidence has been with us for decades; it hasn't made much of a difference. Climate-denial was the original post-truthism.[18]

The existentialist philosopher Jean-Paul Sartre, in the context of Nazi invasion of France — another historical moment when summoning the will to act required exorbitant resourcefulness (and sacrifice) — wrote of the people's desire for reprieve. An excuse for not acting, at least not yet. Faced with what is actually a far greater (but more deniable) threat, we too are dying for a reprieve.[19] We want an escape from the realities we don't want to face and the obligations we lack the gumption to acknowledge. Climate-denial has been so attractive because it offers a reprieve, an excuse for resisting the urgent call to join the resistance.

The truth is obvious to anyone willing to look, and yet it has taken us decades to arrive — finally — at the point where climate-denial is publicly unrespectable. For a public still hoping for a reprieve, the strident voices of shills and 'contrarians' have till very recently been far more seductive than the sober intonations of scientists. This is, I argue, a problem of the will. If you want to achieve something, you have to be willing to accept — and enact — the means to that end. We have been reluctant to will the end (a sane future, living within safe ecological limits) because we have been reluctant to will the means (especially, a drastic reduction in our energy consumption[20]). And so, absurdly, we have failed to take on — to root out — the deniers.[21] We haven't really wanted to believe that anthropogenic climate change is happening. We are only now — with the climate chaos of the last few years, with the advent of school climate strikes, with the uprising of Extinction Rebellion (XR) — beginning to wake up from the 'soft' denial which assumes that anthropogenic climate change could be manageable without serious changes to our economy or lifestyles. Over the last few years, with such events as the Californian and Australian bush fires — and now with Covid-19 — we have begun to have a tiny taste of the cataclysm to come, if we don't start to really move, fast. The kind of eco-driven cataclysm that, we are now starting to realise, will come to rich countries too, and not just to those who have borne the brunt of the impacts of so far.

If we don't will the means to ending the rising tide of destruction, then — in effect — we will the end of our world.

The real issue, as Wittgenstein saw clearly, is not whether we are capable of intellectually grasping the problem. The real issue is whether we are willing to face the reality of the future we are headed towards; whether we are willing to really *feel* the horror of our situation; whether we are willing to accept — more, to seek to bring about! — the measures that will be needed if we are to avert disaster. Whether, in sum, we are willing to truly try.

It is for this reason that my main focus in this little book will not be on facts. The facts largely speak for themselves, to anyone willing to listen, and have done for decades. The problem is that we are not sufficiently receptive to the facts because we do not want to be. The facts can't get in to do their work if your capacity for wisdom and care isn't open. The way I seek to wake us up is by means of an emotionally resonant metaphor and of a little fairly simple philosophical thinking: a simple logic. Logic that a ten year old could master (and, in my experience, they do).

For it turns out that great intellectual acuity isn't required in order for us to change in the way we need to. What is most required is your heart — and your willingness. The question is whether you are prepared to accept the simple logic that I describe in Chapters 2 & 3, chapters that require some philosophical effort of you but no specialist or prior knowledge whatsoever. Then Chapters 4 & 5 outline the kind of needful changes that follow, which make exacting demands of us — but, I argue, they are prerequisites for our continued survival, let alone flourishing.

The first half of this book will explain how our existing values around parenthood and custodianship demand that we care for this living world and work decisively to restrain humanity's destruction of it. The later chapters set out bold recommendations for working collaboratively and deliberatively towards a habitable future for our children, their children and all our many descendants.

We don't want to know

I said I would not make our dire predicament the main focus of this book. But for those who are less familiar with climate science (and if you are feel free to skip this section), I'll briefly rehearse the things you don't want to hear — but in many cases probably already sense, at least deep down — that provide the backdrop for our story. I must lay out the abject direness of our predicament, which is very likely far worse than you've been told.

I will focus primarily on the climate crisis and the interlinked crisis often euphemistically termed that of 'biodiversity loss'. For the climate crisis, that hogs most of the attention, is not all that really matters: far from it. We are at the same time bringing about an extinction crisis, and the primary driver of that so far is not climate damage but ecosystem destruction: the extirpation of wildlife habitats. We are not only recklessly mining the Earth, we are also 'mining' the soil, the fish and the whales. We treat natural 'resources' as though we can deplete or even exhaust them without consequence. Insects are facing potential Armageddon[22] (this is profoundly worrying especially because of their role in pollination, without which our food systems will collapse). And this broad ecological crisis cannot be separated out from the climate crisis. Consider for instance the Amazon rainforest: if it does not retain its biological and ecological integrity and turns to savannah (or even desert), then this cataclysm for biodiversity is simultaneously a cataclysm for our climate system.

The climate-and-ecological emergency is the mega-crisis of our times:

- it increasingly threatens us *now or at least soon*; it threatens to take away the futures — the lives — of those who are living (especially the young);
  and, furthermore,
- unless the transformation we effect is rapid and deep, the threat to our civilisation may well be *terminal*.

Climate itself then is only the canary in the coalmine. Unless we stop mining and digging, then dangerously many other birds too in time will go silent, one sad springtime to come.

For *none* of the ecological crises we're causing can be adequately addressed from within our current paradigm of politics and economics. They can be seriously tackled only if we are willing to make big changes to our system. That is, to the way that, as a society, we live (especially we in 'the Global North', in the so-called 'developed' world). We must be willing to seriously reduce our impact on our home, so as to protect us all. What is called for is a collectively self-protective contraction of the economy; a reduction in the rampant economic growth that our economies have been taught to be addicted to. The recent serious slowdown of economic activity to prevent the spread of coronavirus offers a fragile hope that we may yet be willing and able to take the drastic measures required when a global catastrophe threatens us all. But the way we are so far building back aggressively from the virus is, tragically, tending to fuel the underlying eco-emergency from which it sprang.

And so, there is a spectre haunting our society, our common future: the spectre of climatic cataclysm. Why 'spectre'? Because, in the way I described at the opening of this introductory chapter, it *seems* unreal to us. An air of unreality hangs pervasively over our situation. If human-induced climate change were really as bad as all that, then we'd be doing something truly serious about it already... right? If the situation were really as alarming as Extinction Rebellion and Greta Thunberg make out, then surely governments would act? So, because governments are not acting as if this is an emergency, it seems to follow that the situation cannot be that alarming. Can it *really* be that we are on the verge of committing human civilisation to oblivion? Surely it would take a true 'black swan' event, something utterly unexpected, to accomplish that? Surely if we can see catastrophe coming, then, as rational animals, we'd already be acting to stop it... right?

At this point you might well be thinking, 'But surely we *are* doing something serious; surely governments are in any case acting; that is what the world's leaders agreed at the Paris climate conference in 2015.' It was indeed extraordinary that world leaders managed to agree on anything climate-related at all at the Paris summit, after the debacle of Copenhagen in 2009. It was, perhaps, as good an outcome as could be expected: because every country in the world had to agree, in order for the talks to work. And they did.

Paris was an extraordinary diplomatic achievement, and, realistically, it is hard to hope for much more. This makes the truth about the accord harder to bear: the much-lauded Paris Agreement on climate is a paper tiger. It's dead on arrival. Even if the Paris commitments that countries have made were achieved *in full*, it would not be enough to stop dangerous climate change.[23] It would, in fact, probably result in around three to four degrees of global overheating, triple what we have so far. That would be enough to demolish our civilisation.[24] Enough, over time, to raise sea levels by twenty-five metres and, much sooner than that, to complete the job that Bolsonaro has begun, of burning the Amazon rainforest (which seeds much of the world's rainfall, and some of its oxygen), turning it into savannah and possibly in time desert. A three to four degree rise would accelerate the catastrophic 'feedbacks' already in process, causing even greater ice-melt as the poles turn less white (and so reflect less sunlight away), and quite possibly triggering the massive release of the highly potent greenhouse gas methane (much of which is stored near the poles). Terrifyingly, we may already be quite close to unleashing this 'methane dragon'; in fact, this process might even be said to have already begun.[25] If more than a *fraction* of the methane stored under ice gets released, there will be a runaway heating effect which will probably wipe out most of humanity.

If we limit ourselves to achieving what Paris commits us to, then we are almost certainly committing ourselves to the collapse of civilisation as we know it. To say the matter plain: even if the commitments obtained under the Paris process are achieved, climate devastation will still almost certainly bring down civilisation as we know it. Paris achieved what was politically possible, not what is needed.

But it's worse than that. We can say with near-certainty that, barring an unprecedented change in consciousness, the parameters set out in the Paris Agreement will *not* be achieved. The treaty is non-binding, and virtually every country in the world has plans (for road-building, for air-travel expansion, for ramping up intensive animal agriculture, and so forth) that contradict their Paris commitments. Some of these infrastructure projects continued even during the Covid-19 lockdown, notably Britain's carbon-heavy, ancient-woodland-demolishing, high-speed rail system, HS2. Lockdown had its environmental silver linings, of course: we saw fast falls in pollution, including in climate-deadly carbon emissions, during the height of the pandemic. The challenge now is to harness the public's acceptance of the need for economy-limiting measures in times of existential threat to argue for levels and forms of economic activity that are long-term ecologically survivable.

I put it to you that, deep down, you *know* that the path we are on, or even a 'reformed', 'improved' version of it, is a high road to cataclysm. We will not even meet the toothless demands of the Paris Agreement. You know full well that endlessly building more high-tech transport infrastructure (starting with airports) points in the opposite direction to reducing our burning of fossil fuels. You know that the endless 'growth' of the economy is at the cost of the ecology.[26] ('It's the ecology, stupid...') But it is hard to face this. The real difficulty is in allowing what you know in your bones to come to full consciousness.

Because, once it does, it cannot be un-known.

And here is something you may well *not* know, the nail in the coffin of Paris's credibility as a plan for saving humanity. The hopeful scenarios of the Paris Agreement depend upon the availability of magical-sounding 'Negative Emissions Technologies' (NETs) to suck carbon out of the atmosphere (so that in the future we would allegedly be able to have less-than-zero net carbon emissions).[27] These are technologies that *do not exist* and that, even if they did exist, would be reckless in the extreme to deploy (as I shall explain in Chapter 4). This is the dirty secret of Paris, and the dirty secret of the UK Government's 2050 net zero target: the word 'net' scoops up a multitude of sins. Sins that are being indulged now, as we gamble our children's very future on non-existent NETs that are most unlikely to save them.

There will likely be no net to catch our kids if and when ecosystems (and, as a consequence, entire economies) start to cave in on themselves. No one wants to think about this, but the harsh reality is that we are already not far from making it virtually impossible for human civilisation to outlive the century. We are smoothly marching over a cliff, pushing our most vulnerable in front of us as we do. We are gambling the entire human future, without a backstop. A climate-devastated future is not a 'black swan' possibility; it is not some surprising, unexpected event: it is a *white* swan.[28] It will come and overwhelm us — unless we change course far more radically than is dreamt of in Paris's philosophy.

That future is already here, for Bangladeshis, for Pacific Islanders — even, in the record-breaking summer of 2018 (since when we have had many more such records broken), for many residents of Houston and Florida and parts of Greece. That future arrived, in early 2020, for many Australians (including a billion Australian animals). Their fate prefigures that of our children, grandchildren and great-grandchildren, the multiple holocausts to come. The summer of 2020 was horrifically, unprecedentedly bad, in terms of wildfires (including in the Amazon, again, and California), crazy temperatures in the Arctic, the collapse of ice-shelves.[29] But much of this wasn't even noticed, as the world was gripped with pandemic-fever.

It has been shown beyond reasonable doubt that anything remotely like a reformed business-as-usual path puts us on course for climate-nemesis. This dire outcome is, quite simply, what anyone with a basic understanding of the situation should now expect.

It is true that there are still some grey areas which could — *could* — turn out to work in our favour. We don't know the exact 'climate-sensitivity' of the Earth system: we don't know exactly how sensitive it is to the carbon-pollution we are flooding it with. We don't know all the feedbacks that are likely to kick in, nor just how bad most of them will be. There might even, if we are very lucky, be as-yet-unknown feedbacks that will actually buy us time. So, we don't know how long we've got. But none of this means that we can relax. Far from it.

For crucially, these uncertainties, as I shall detail in Chapter 4, underscore the case for radical precautionary action on climate: for uncertainty cuts both ways. This is what 'climate sceptics' deliberately forget: that for every uncertainty that might mean things will be less bad than we fear, we are also exposed to things being potentially even worse than we fear. The grey feathers in the white swan's plumage change the situation not one bit — except to underscore how we not only have a (broadly) predictable catastrophe facing us, but, furthermore, one that may exceed most of our models and even our imaginations.[30] It is beyond reasonable doubt that we are driving ourselves and our loved ones towards the edge of a precipice; maybe one with a fatally larger drop below it even than our best current science suggests.

Catastrophic climate change is a white swan; and even the odd grey or black feather only underscores the precarity and unpredictability of our current situation. The gravity of our exposure to incalculable harm.

A way to not turn away?

The situation outlined above raises deep ethical, philosophical and political questions. Firstly: how can we look our children in the eye while continuing to allow this cataclysm in the making? (Maybe this is why we typically *don't* quite look our children square in the eye, on this determinative issue: why virtually all of us, and not just those on the nastier and stupider fringes, are — in practice, most of the time — in some form or another of climate-denial.)

Secondly: how can we be woken up? You may have heard of frogs who will doze till they die in water that is gradually heated up. But here's something that you probably didn't know, a piece of potential good news: actually, most frogs *jump out* and save themselves before the temperature gets too high.[31] Is it not reasonable to hope that we can be at least as rational as frogs? How do we learn to act as 'wise frogs'? How do we jump out of the saucepan before we boil ourselves alive? It seems to me that we do not jump because we do not want to acknowledge the problem (and to acknowledge our responsibilities: to acknowledge those for whose fate we are responsible), which is startling given the way it stares us in the face. To deny the obvious takes much more effort than to deny the unclear (and this of course is why for years now climate-deniers have been trying to argue that the evidence is *not* clear). And yet, I say again: virtually all of us are complicit with such denial, most of the time...

How can we call ourselves *rational* animals, as the philosopher Aristotle claimed we are, given all this? The truth is that we can't, as long as we continue to value our desire for reprieve over a humble acknowledgement of the facts. That is why I assert that, uncomfortable as it may be to acknowledge it, it seems that the majority of us have much more in common with climate denialists than we like to think.

I'm going to try to change that.

The book you are holding in your hands aims to provide a cogent way of thinking about all this. A way to not look away. The enormity of the facts and the urgency of the situation can be overwhelming, I get that. I'm as daunted as you are.

The temptation to turn away is profound: it's not your problem; these disasters are happening to other people, elsewhere in space and time; maybe it might not be as bad as 'they' say. But these are strategems for avoidance, and little more. I want to share with you how, starting from uncontroversial propositions about the love and care you feel for your own children, we arrive at a compelling case for taking action to prevent the destruction of ecosystems and build a liveable future for billions of as-yet-unborn strangers. I mean: a case that you will actually feel compelled by. Even if you *think* that you don't care about the planet, or even not about other people beyond your own family.

Clear thinking, combined simply with your genuine care for your own kids, really can do the trick of awakening you — and, in the same way, everyone else.

If at this point you feel desperate, or frustrated, wanting already to *do* something, then do it. Stop, put the book down for a minute; and (say) switch to a 100 per cent renewable electricity supplier such as Good Energy or Ecotricity. It doesn't take long. Or if you have already done that, then get government help to get your home better insulated or fitted out with solar hot water.[32] That takes rather longer but can be even more significant. Or at least maybe switch your browser to the wonderful tree-planting site, Ecosia, which plants trees as you internet-search.

But do not be under any illusion that taking such conscience-salving steps amounts to more than the first tiny gesture in the direction we need to be going in. There is no individualistic solution to the ecological emergency. This book seeks to point you in the direction of travel that we will take together, if we choose not to fail our children. The journey through the following chapters requires a much more substantial level of courage and ambition than any green consumerism. It needs you as a citizen. And, in fact, as a whole human being. Beginning with the meaning of your role as a carer.

Let me close this introductory chapter by sketching how the remainder of my book, through four chapters of simple, accessible philosophy — combined with a few imaginative exercises and the occasional example from films, books and art — can enlighten and encourage. How, from premises that virtually everyone agrees to, we reach an extraordinarily salvational conclusion. How we might yet snatch a kind of triumph from the jaws of defeat, if we are brave and realistic enough to see, and *to try*.

Chapter 2 explains why caring for your kids means you'll care for the whole human future. It begins from the observation that facts (taken by themselves) have failed us. Scientists thought that sharing the facts about the climate and ecological emergency with us would be enough. They were wrong. Instead, I begin by focussing on what we — on what *you* — most *value*. And what do humans value more than their own children? So, let's make the conservative assumption that your children are what you most value.

Next, I argue that concern for your own children isn't real if it doesn't include the same level of concern for *their* children too. This is because they love their children more than anything, and you don't love them if you destroy what they love. This logic iterates: if you love your children, you are committed to loving all your/their descendants. The only non-reckless attitude to have toward the future of the human race is thus to care for it all because over time your descendants will, so far as you know, be spread anywhere or everywhere in the world, marrying away like rabbits. Real care just for your own children, therefore, entails the same care for the distant future of the whole human race. (And similar considerations apply *vis-à-vis* the young who you care about even if you are yourself childless.) This argument implies that it is useless to try to take care of your kids by (for example) building them a bunker, or making them incredibly wealthy. Although this might well work for a few decades or even generations, it won't work for your distant descendants if our collective life-support systems are breaking down. (This chapter also explains why the logic I have set out works for the parenting of the future undertaken by all of us, including those of us who don't have kids of our own.)

Chapter 3 describes why caring for the human future means you care for the planetary future. Many people, even when convinced by the logical argument contained in Chapter 2, say that they only care about humans, and so still don't see any strong case for taking care of our ecosystems. But, I argue, anthropocentrism (the placing of humans first) *equates* to ecocentrism (the placing of the planetary ecosystem first). This is because, in the long term, the only non-reckless attitude to have toward the human race is to safeguard its (our) continued existence by protecting the ecosystems on which we depend. We depend on them utterly: for everything from regulation of the atmosphere (of pollution, of weather patterns), through new medical cures and foods, to places to seek refuge and solace and rest. It makes absolutely no sense to think of replacing those — or putting them at existential risk (risk of no longer existing). We are *nothing* without a living planet. We are nothing but part of it.

This inescapable fact turns concern for humans (a concern that, as Chapter 2 shows, extends into the distant future) into concern for the planetary ecosystem reaching into the distant future. Even if you (think you) don't give a fig about fluffy animals and trees, it nevertheless turns out that you *do*, because the only rational stance to take is to assume that, in the long-term, they are essential *for humanity*.

By the end of Chapter 3, the book will have established several key points. Chapter 2 turns concern for your own kids into concern for all humans, extending deep into the future. Chapter 3 turns concern for humans projected into the deep future into concern for the planet projected into the deep future. Together, these two chapters take one from loving concern for one's own kids only into loving concern for the very-long-term future of planet Earth as a living ecosystem. Given the grave threat hanging over the planetary future, we can start to see where that loving concern needs to be directed.

Chapter 4 starts from the premise established in the foregoing sections: that just caring about your own kids entails deep long-termism, ecology for keeps. Given the predicament already outlined here in Chapter 1, this requires a revolution in how we organise ourselves. But what should this mean in practice? How can this revolution be implemented? We need bold new institutional mechanisms for embedding — and enforcing — this revolutionary change of attitude. My three core proposals are as follows:

1. *Citizens' Assemblies*, constitutionally empowered to take the necessary decisions to bring us from incipient disaster to survival and flourishing. 'Representative democracy' has failed us; it has put no meaningful obstacles in the way of our 'progress' toward nemesis, and thus badly needs serious supplement. Representative democracy as we know it has failed to represent the interests of future people (and other beings). We need to find a new way of representing the people. We need a designated body to undertake the vital work of deciding how exactly we make the difficult transition into a future that will not kill us. Politicians should 'outsource' this work to assemblies drawn from the citizenry, on a geo-demographically representative ba-

sis. (Smart politicians will understand that doing this protects them from taking all the heat for the tough decisions such bodies will arrive at.) The citizens in these assemblies, unlike parliamentarians, will be mostly drawn from state schools, will be half women, will include the young in proportion to their numbers in the population; and so forth across ethnicities, occupational groups and more. These assemblies will be presented with the dark truth about our ecological predicament and, advised by a wide range of the best experts, will deliberate about how to save our common future. Immune to the long arm of the lobbyist and to the vicissitudes of opinion constantly mediated by the press and by nefarious social media algorithms, outside the swirl of influences that have largely neutered those politicians who desire to do the right thing (yes, they do exist!), they'll decide a path forward that will swiftly reduce our deadly greenhouse gas emissions to zero and rapidly curtail our devastation of biodiversity (i.e. of life).

2. *Guardians For Future Generations.* Once the Citizens' Assemblies have done their job of overseeing the transition — of dealing with the pressing urgency of the situation — then a new permanent arrangement is also called for. We need radical reform to our system, reform that will guarantee a long-term vision. Sitting above parliament, there should be a permanent panel of Guardians for Future Generations: like the Citizens' Assembly, this would be a kind of 'super-jury' which can radically rein in the short-termism of 'liberal democracy'. These juries — like all juries — would be highly democratic, with their members picked by lot rather than by vote. Guardians would set us on the road to an enriched democracy that would, in effect, *include* our kids *and* their kids and so on, for the first time ever. Finally, we would start to make decisions to act in the long-term interests of people and planet.

My third proposal is to adopt a philosophical, ethical and legal framework that would inform the decision-making of the Citizens' Assembly and Guardians for Future Generations. The keystone of this framework:

3. *The Precautionary Principle*, which to date has been honoured more in the breach than in the observance. The Precautionary Principle is embodied in the common-sense sayings 'better safe than sorry' and 'look before you leap'. It states that when you lack full evidence and potential consequences are grave, you need to err on the side of taking care. It doesn't urge us not to take risks: it doesn't say, be so chicken as to refuse to walk across the road to get to the other side. It does say: don't walk across the road blindfolded. And if you really must cross the road blindfolded, then do not under any circumstances drag your children along behind you as you do. In this way, this principle enjoins us to err on the side of safety, rather than wait for confirmed evidence of threats to our very existence. For, by the time all the evidence is in, it may well be too late. This principle also acts therefore as a potent tool against 'climate sceptics' who, stupidly and immorally, would gamble on the overwhelming evidence for climate science turning out to have been wrong, and would gamble on all the remaining uncertainties in the science breaking our — humanity's — way. Our current way of life *is* a gigantic leap without looking. It is utterly reckless; as would be relying on mirrors in space or other sticking plasters to save us from the overheating of our atmosphere that we have set in tow. It imperils the very future of humanity. Constitutionally embedding and acting consistently upon the Precautionary Principle would revolutionise our politics, economy and society. It would ask us to act as if the worst case scenario is probable until we know for sure that we have stopped it from unfolding. And the good news is that we now

have a precedent in those East Asian and Antipodean countries that acted effectively, precautiously, to suppress the coronavirus, and which came out better in every way than those American and European countries that acted slowly and recklessly.

Citizens' Assemblies, Guardians for Future Generations, and the Precautionary Principle: radical changes such as these, bold and immoderate as they might sound, represent the bare minimum that will be required in order to put us on the kind of path that my argument in Chapters 1, 2 & 3 logically entails. Fighting for them, or ideas like them, and fast, is therefore the very *least* we need to do.

You cannot say you did not know. All that remains is to consider what *you* will now do.

Chapter 5 examines what it might be morally incumbent upon each of us to do to help drag the climate back from the brink of catastrophe. In the corona crisis, we sought to care collectively for our elders. In the longer climate and ecology emergency, it's time for the favour to be returned: it's time to take care of our kids, and their kids, and so on. For that, we need what I like to call *wholescale change*. Our chances of successfully transforming our societies in the ways I have outlined may be slim. But we must try. Let's put the whole of ourselves into this effort to scale up our response to meet this challenge of the ages. Giving up would mean giving up on your kids. It would mean that your love for them is just an idea, but not something you are willing to act on.

But, of course, you mean it: you *do* love them. And so (of course) you are determined to do the right thing and make things better. There's nothing more fundamental or beautiful than our concern for our children. So, there's nothing that can stand in our way once we decide to act on that concern. We have already begun to see the awesome power of this concern with the rise, side by side and hand-in-hand, of the school climate strikes and Extinction Rebellion. These movements promote honesty about the gravity of the problem and speak with clarity of the fact that we have little left to lose. They have drawn the consequences and risen up.

What action is called for from us in order to bring about the revolution we need? To *institute* the new ideas set out in Chapter 4? Here's the thing: it's no good saying, 'Well, "they" ought to do such and such.' It's too late hope for 'them' to ride to the rescue. The change has to start in this moment. Here and now. With us. Which, right here and now, means with *you*. With the you who is reading these pages. My pitch then will be: your money and/or your life. You need, at minimum, to devote either your time or the bulk of your financial resources to this cause: the cause of changing the world so that there is a long-term healthy planetary future, in which all (y)our descendants can flourish. And you need to think about what is credibly a way of realising this cause. Giving your money or your time to (say) the National Trust, or the World Wildlife Fund, is not. Giving it to a movement more like Extinction Rebellion, or to a similarly bold movement with even bigger reach, might be.

Such giving is what really caring about your children now means... Giving your all to mitigating the eco-emergency, to safeguarding your children's future. This is how truly caring for your own kids can unleash a political and personal energy that will propel the fundamental changes our civilisation needs, to transform and survive. In the relatively short compass of the pages to follow, I will demonstrate how relatively modest, common-sense thinking leads us from the simple proposition that everyone wants the best for their children to the radical reappraisal of our society and economy.

And the rest will be up to you.

The book ends with *A Proposal*. For a movement that doesn't exist yet, but that could go far further than those which already do, toward the goal of adequately parenting the future. The book ends, then, with an invitation to you: to create momentum for the birthing of *this* movement. Parents For A Future...

At the very foundation of the essay as a genre in the English language, Francis Bacon recognised that an essay necessarily tries — tests — the abilities of both writer and reader. An essay is a kind of experiment, begun by an author. But the experiment is necessarily incomplete. It needs to be completed *by the reader*.

Essays — unlike textbooks, or propaganda — demand the reader's active involvement, in order to be fulfilled. An essay is not merely a palimpsest; but nor is it something simply to be swallowed as is, passively accepted. This little book aims to put a possibility out there; and then requires readers to step up to the challenge. If this book is completed, it is at any rate not complete when you reach the final page. It will be completed by what you — and many others — choose thereupon to *do*. For *that's* how you really put my work to the test.

I hope you will. And so do your kids.

2

Truly caring about your own children is enough to save the whole human future

'Human nature is such that it cannot be indifferent even to the most remote epoch which may eventually affect our species.'
– Immanuel Kant[33]

Three numbers to know

To set the scene, I want to make sure, reader, that you have taken in just how appalling our current situation is. I'll do so by way of three simple but profound statistics, three numbers which need to be much, much better known.

First off, consider that the human race may be driving species extinct at the rate of about one *every ten to fifteen minutes*.[34] Whereas the normal 'background' rate of natural extinction may be as low as about one a year. By the time you have finished reading this book, even if you read it at one sitting, about ten or so species will have been killed off, forever, by us.

Pause a moment, please, to let that jaw-dropping number sink in.

And next, consider just how deadly the fossil fuels that we are wrecking our atmosphere with are. Consider what happens when you burn petrol in your fuel tank. Intense heat is produced to power the engine. Carbon is also released. It goes into the atmosphere. It stays there for decades, even centuries.[35] (This is one reason why true long-termism is demanded by the climate crisis; there are big time-lags between when we release CO_2 and when it does its worst damage. Even if we were to stop all carbon emissions tomorrow, the damage we have

go on getting worse for some decades to come; carbon
atmosphere for a long time, sets off vicious feedback loops
mages the oceans, which absorb excess carbon and heat.)
our common predicament is so extremely severe; and why,
as I enjoin in this book, we urgently need to learn to look ahead.

Once it's in the atmosphere, generating the greenhouse effect,
how much overheating of our Earth does CO2 cause? That carbon
released from burning a litre of petrol, do you think it traps perhaps
the same amount of heat in the atmosphere as was released by burning
the petrol in the first place? That would be extremely worrying, if the
greenhouse effect doubled the warming effect of burning these fossil
fuels, this vast reserve of carbon that has until recent decades been
buried, since times when the Earth was way hotter than it is now. Or
maybe it's worse than that: twice as much, perhaps?

Ten times as much?

The actual figure? Carbon traps sixty thousand times as much
heat, over its 'lifetime', as was produced in the fire which released it.
Put another way: when you burn *one* litre of petrol in your engine and
release its carbon to the atmosphere, it's as if you start to release the
heat of burning *sixty thousand* litres of petrol.

Do you see, now? Fossil fuels are Weapons of Mass *Self* Destruction.
Without doubt, they will destroy us, they will lead to our civilisation
collapsing, unless we move swiftly and surely to stop releasing them.
And our children will have the worst of it. If, reader, you are of roughly
a pensionable age, then you might just get away with having a good
life before the deluge (though don't bet on your last years being happy
ones if they are spent on an exponentially heating planet). But our children
and grandchildren *will* suffer the results of the deadly greenhouse
pollution that we bring about today and tomorrow. And unless by 'the
day after tomorrow' we've seen sense and changed everything, then
their lives will be poor, nasty, brutish — and short.

The situation I have just sketched concentrates the mind (and the heart). True, the mind repels such numbers: they are almost too awful to know. But, equally, they are too important, too transformative, *not* to know, not to allow yourself to take in. They change everything.

In this chapter, I ask you to heed the sense and necessity of a *deep* care for our children. I argue that the care we profess to have for our kids means nothing if it is not properly long-termist. To be precise: if you care deeply for your children (as you surely do), then caring for your own descendants becomes — through logic alone, through you acting in line with what you already believe and feel — a care that reaches into the deep future. Because caring for the next generation (your children) transfers to the subsequent generation (*their* children). And this process repeats, forever. This care of yours, therefore, cannot safely be geographically restricted because one's own future generations will most probably gradually disseminate across the globe. What our own children turn out to mean to us, then, is nothing less than a deep (into the distant future) and wide (across the globe) loving care.

And the third of the three numbers I promised you? Thirty. That is the number of years for which the UK Government is planning to go on making this existential emergency worse. They promise (if you can believe them) to bring carbon emissions down to zero by 2050.[36] So, we have an emergency threatening our collective futures most direly, especially the futures of those twenty, thirty, forty years younger than us, and what does our Government's 'world-leading' plan intend to do about it? To stop making it worse — *in thirty years' time*.

Instead, it is time to get serious about what it really means when we say that we love our children.

What *do* people value?: our own

Let's make a very conservative assumption, one so minimal that it cannot be accused of being overly altruistic, as, quite often, we 'do-gooders' are accused of being (Isn't it 'funny', how even doing good gets demonised...). Say you *only* care about your own family, your own kids, and not about other humans. Doesn't everyone, from the most idealistic to the most cynical, concur in profoundly valuing at the very least their own children? This is what many people — including those who are seemingly quite uninterested in 'the environment' or the distant future — say, when quizzed: that what matters to them is their *own* family, and particularly their children; that they would do anything for them. Let's see what follows from such care.

What does valuing your own children so deeply really mean? Well, if you value your children — not just as playthings for you or 'carbon-copies' of you, but for their own sake — then it follows that you'll value what they most deeply value. This doesn't mean that you must value what they superficially appear to want. Nor does it mean that you have to support their desire to engage in behaviour that is self-harming or endangers their wellbeing. But if you claimed to value your children for their own sake, *while systematically stomping on what they most deeply valued* — not only throughout their childhood, but throughout their life — your claim would be obviously suspect.

So, loving your children means taking seriously the things that they care about. And what can we expect our own children to value deeply? At first sight, a difficult question that might even seem unanswerable. In beginning to answer it, it seems reasonable to expect that, at least at the level of fundamentals, our children will not be completely different to us. The one thing we can surmise without much likelihood of going wrong, then, is that they will deeply value *their own children*. And already perhaps you can see the crux of my argument, my metaphor, in all its simplicity. For the point that I have just made iterates endlessly down the centuries and millennia, like a mathematical induction. 'Merely' valuing profoundly one's own children rapidly ramifies into valuing just as profoundly all one's descendants. And your descendants ad-mix, the longer we look into the future, with more in more of the human race. In the very long run, we're all one family. With a deep enough temporal gaze, all the world's people are as good as being your children. Care for your own children, then, *equates* to a wider care for all people in the distant future.

The line of thinking that I'm outlining has the advantage of not requiring one to value the distant future directly for its own sake. It is, I would suggest, far more emotionally resonant than direct appeals to the needs of abstract 'future generations'. *All that my argument requires is that you really do love your own kids.*

Now you, reader, will doubtless be a decent and charitable person who cares about other people's children a lot too. But the beauty of my idea is that it doesn't *depend* even on such charity and decency. It depends instead on values that are widespread and uncontroversial. My argument, rather than requiring immediate assent to ecological or ethical principles, is simply directed to any and all parents, including those who are, quietly or openly, far narrower in their concerns. Have you ever wondered if your focus on your own kids' life chances is somehow a bit narrow, or selfish? Do you ever feel like you're a bit of a 'helicopter mom' (or dad) who only cares if their own child succeeds? Maybe it worries you sometimes that you just can't care about other people, other *strangers*, the way you care about your family. Well, if so, my argument is for you. If it works at all, then it can work for virtually everyone.

From modest premises to massive conclusions

I have established, then, that caring for one's children means caring for all one's descendants. But how can one determine *who* exactly to care for, as the generations become more distant and the picture grows hazier? It's obvious that one can't with any exactitude conceive of *who* one's descendants will be because they haven't been (ahem) conceived yet.

In getting at least a basic sense of who they will be — who they will be relative to oneself, that is — one can make a start by thinking of family-trees. A reasonable baseline assumption is that, just as one has more and more ancestors as one goes back in time (twice as many grandparents as parents, and so on), so one will probably have more and more descendants as one goes forward into the future. (Think of those heart-warming stories about — or images of — very old people surrounded by their three children, their eight grandchildren, their fifteen great-grandchildren, and so on.)

Over time, these descendants are likely to be dispersed more and more widely across the world. This process may be very slow; I'd like to see a future in which we travel less, are less hyper-mobile, grow better roots in our localities. But even if we one day move at the speed of sail again, rather than that of jets, we would, at least, be wise not to assume that our descendants will all live where we do. So, truly caring for your kids involves caring for all your descendants, and now we see that those descendants are most likely more numerous and (over enough time) more geographically widespread with every generation. Of course, they *might* not be; lines of descent do sometimes extinguish, for example. But it would be very foolish to *assume* that this will happen. Given that you cannot anticipate exactly where your descendants will live or end up, the only sensible — precautionary — course is to assume that there will be no geographical limits to their spread, given enough passage of time. Specifically, in the context of the contemporary globalised consciousness, it is obviously even more foolish to *assume* that all your descendants will live in the same country that you do. This happily overcomes the risk of parochialism that might have been thought to follow from focussing one's care on one's own kids. It also undercuts the notion that one can protect one's descendants by creating one very secure place (I will return to develop this point).

Starting from the unobjectionable minimal assumption that we deeply value at least our own kids, we can now see that what we deeply value is not only our own kids, *narrowly* conceived, but the ongoing un-folding of the descending generations to follow us (and them). In other words: care for the latter is an unfolding of care for the former. Care for one's own children *amounts to* care for virtually the entire human future. It encompasses the most remote epochs that we can envision or plan for. (See my epigraph, from the philosopher Immanuel Kant: this chapter constitutes a way of realising Kant's point, which hitherto may have seemed to require something excessive and unreasonable of us.)

I am not of course claiming that our kids are *all* we value. I am not saying that we must sacrifice everything for them and for future gener-ations (though some parents might say that, and the last thing I would want to do is belittle such marvellous altruistic sentiment). I am com-mitted only to assuming that our care for our descendants is non-nego-tiable — an assumption that is reasonable and uncontroversial. From this modest premise, we can deduce the strong conclusion I am aiming at in this chapter. Care for the human future out into deep time.

But, it might be countered, what if the population of the Earth massively decreases?[37] This seems (on current trends) possible, if not probable; either as a result of a growing preference for having fewer children, targeted population-reduction policies (e.g. education of girls, free family-planning), and environmental consciousness of the Earth's limited resources; or as a result of uncontrolled collapse...

Say for several generations each family were to have on average one child, so that the population halved, and then halved again: would this defeat my argument? Would it show that the future gradually became less important, the further from us it is? I think not; because in that scenario, we would notice more easily that what matters is each succes-sive generation, equivalent to the last. One child would be the focus of your love, likely all the more precious for being the one and only. (Only don't spoil them!) And so to your one grandchild. And beyond.

And of course, as I've already implied, the very reason for such population-reduction would be likely to include a concern about the *self-destructiveness* (when summed across the planet) of having as large a human population as we currently do, let alone a larger one. We would presumably be reducing our population voluntarily to prevent the involuntary reduction consequent upon collapse. It might sound paradoxical, but care for one's child(ren) — which, we have already shown, is equivalent to care for the whole distant human future — can be expressed through having fewer of them.

Either way, caring about the future remains undiminished and undimmed over time, provided that one truly does care just for one's own child(ren).

Childless?

I must, of course, consider the obvious question that has probably been forming in the reader's mind: 'How is my argument relevant to the entirely childless? Or is it irrelevant to them?' In this case, it would seem to have limited power. However, the first point to make here is that the scale of this objection is not that great, for the childless are a relatively small minority. If my argument is limited by 'only' working for eighty to ninety per cent or so of humans, then I'll remain pretty content.

Let's also remark that it turns out to be no objection against my argument to note that, historically, many childless people have been childless because they are gay/lesbian.[38] Arguments like mine which emphasise having children might be experienced by some as running the risk of adding to anti-gay prejudices. My main response to this worry is outlined in points (i) and (ii), below. But let me say up front that, at least in a growing number of countries such as Australia, Canada, the USA and the UK, this point is in any case, thankfully, less and less empirically true. Non-heterosexual people who want children are increasingly able to have them, through one means or another.

Nevertheless, the question of whether the existence of childless people poses a problem for my argument deserves patient attention. By examining the various reasons for childlessness, I believe we can come to put aside any concerns that it might represent a serious problem for me. I'll start then by saying that the childless can be divided roughly into the following three categories:

i. *Those who are childless because they are motivated principally by care directly for others.* Those whose lifework is substantially motivated by caring directly for future generations — including those who decided not to have children so as to be able to devote themselves to future generations — obviously, simply don't require the argument of the present chapter to motivate them in the first place. Those who are childless because they are, for example, looking after their parents, or manifesting their artistic abilities, writing for the public good or in the public interest, running demanding organizations, working night and day to help save our world, etc., are in many cases motivated indirectly by a care for posterity, and in most other cases, I would suggest, are at the very least unlikely to be hostile to the argument of the present piece. For they (we: I'm one of them) are driven by ethical concerns, and those concerns can be easily seen to have something important in common with the concerns of those motivated directly by *care* for the next generation(s). Indeed, some of us are motivated not to have children directly by the desire to make the world in general, and rich countries (with heavy ecological footprints) in particular, less over-populated — and so better for all the children who *are* here. This group also includes those on 'birthstrike': those who, though wanting children, have made the difficult decision not to have any, because they don't want to bring children into a world so devastated as ours (at least not until we cease the devastation).

ii.   *Those who have nephews and nieces etc., about whom they care profoundly.* This category may overlap with the first. Into this category fall many of the childless-by-choice (including, again, myself) and involuntarily childless, many of whom lavish on their nephews/nieces the love that they regret not being able to lavish on kids of their own. (Of course, some of these childless-not-by-choice adopt, and thereby re-enter the main argument of this chapter.) The category of the involuntarily childless includes both those physically unable to have kids, and those who have not (yet) found a partner to bring up a child with and do not want the work and responsibility of bringing up a child alone. Such people will typically seek an outlet for their care, either through helping relatives' or friends' children, or through investing themselves in organisations such as Save the Children. They will therefore, I assume, be open to a simple extension of my argument. (Indeed, one might go further. Such people often manifest a kind of 'collaborative parenting' that could — I'd suggest — be a model for how we should envision our common future. If the argument of this book works, then it turns out that we are all in a way collaboratively 'parenting' the future, including the distant future. Perhaps we should more consciously adopt this way of thinking about our common enterprise of care...)

iii.   *Those, if any, who chose not to have children because they are essentially short-termist and hedonistically selfish*, or who, not having children, have decided self-consciously to be thoroughly short-termist about their lives. This is at most a very tiny minority indeed. And one that can be considered the exception that 'proves' the rule. One need not have anything to say to them, any more than one need have anything to say to complete cynics or moral nihilists.

I am not, of course, arguing that having children is necessarily good — still less that having more is better. On the contrary, many of those in the first category are likely to be especially admirable. I am not at all arguing, 'Let's have (loads of) children; this is the best thing of all'. I am rather starting from the fact that most people want to have children, professing earnestly and (I trust) sincerely to love their children most deeply of all. I am working from the fact that it is essential that this desire not vanish, unless one wants the entire still splendid human adventure to come to a halt, with the population shrinking painfully to nothing, without reproduction. The undesirability of that kind of scenario is brilliantly investigated in Alfonso Cuaron's extraordinary film *Children of Men* (2006),[39] which examines how much of the meaning of life and our sense of hope would diminish in a world without reproduction.

Rather, I am developing an argument which does not rely on the minority of ethically concerned consciously childless people bearing the whole responsibility of saving our future. It may be that some who follow my argument will reach the conclusion that they can best care for the future by remaining childless. That's the conclusion that I myself reached many years ago. And, as I've already suggested, in touching on the scale of the human population, that would in fact be a healthy conclusion for many more of us to come to. But my argument delivers hope, in that it does not rely on people who are willing to remain childless, a group which seems likely to remain a minority. On the contrary, my case is directed first and foremost to the vast majority.

The future lasts a long time

I'll now consider another possible objection: does my argument require us to suppose, very implausibly, that the future will be literally infinite?

I argued above that (y)our care for (y)our kids carries over endlessly into the future: by loving them, you are committed also to supporting them in their love for *their* kids. And on and on. But it is important not to be distracted by taking the idea of 'endless' iteration too literally. People might think, 'At some point this will *all* end (the Sun will turn into a red giant and swallow the Earth; eventually the universe will suffer heat death), so value can't simply reside in the eternal perpetuation of humanity.' This is partly right: the value *isn't* in the perpetuation; the value is in every step along the way. The point is that there are indefinitely many steps, not just one (or two, or three), and that each step matters equally. One therefore isn't actually valuing the first 'step' if one values that step only. That is to say: you don't actually love your children if your actions (or inactions) clearly compromise the future of their children. If you care about your kids, then you care about their kids: and so on, in practice endlessly.

The concern might now be raised that, even if my argument doesn't demand that we be able to project into a truly endless future, still it makes implausible demands of our knowledge about the future. For example, in the very distant future humans — our descendants — may be very different from us; they might be 'post-humans' or cyborgs. How can we know how to take action now to facilitate care for these unknown and unknowable future inhabitants of the Earth? My argument, here, might then seem to be asking too much of us. My response to this is that we can — and only need to — direct our care into the future as far as we can foresee a plausible need for it. If 'post-humans' somehow don't require some of the 'ecosystem services' that we do — if they don't need food, water and shelter like we do — then fine. But it would be the height of reckless folly — it would be radically unprecautious, it would be (to put it plain) plain stupid — to gamble everything on this. For such recklessly gambling probably destroys the prospects of our descendants who may or may not ever get to that 'exalted' point.

I think, furthermore, that, to be on the safe side, we should care about the survival of species which might in millions of years' time (if we go extinct or close to it) be capable of taking over from humans as the 'dominant' species[40]: the descendants of bonobo chimpanzees or of elephants, perhaps; social whales and dolphins that might well return to the land; etc.. I believe it possible that they would do a better job of care-taking the world than we will/are. It would be an awesome crime to pre-empt their existence and their chances by terminally wrecking our world. Likewise, and closer to home: it is terribly difficult to foresee closely what people will care about, a few generations from now. There is even currently speculation that we will eventually succeed in transcending reproduction altogether, which would of course fundamentally challenge the terms of my argument. But it would be utterly, abhorrently reckless to bet everything on this outcome.

In sum, my argument doesn't demand excessive foreknowledge of the future. It demands only reasonable deep care for the (reasonably deep) future: a sensible and necessary refusal to gamble on the future not turning out to need such care from us, now.

For the same reason, the unwisely widespread assumption that we'll escape climate catastrophe by an extreme technological fix, such as by 'emigrating' to Mars, is utterly reckless. I suppose we (or rather, an 'exalted' tiny minority) might conceivably one day emigrate to Mars, though I myself consider it very unlikely indeed; furthermore, like nearly all proposed technological fixes, the vast use of resources and energy that would be required to try to move us in that direction would itself speed up ecological breakdown.[41] The point is that the unknowability of the future rules out *assuming* that we will do so, or that we will in any other way magically fix our problems through technology.

Wackily extreme technological optimism must not be used to undermine the enduring, real need to stay safe by protecting our children (and the planet that they rely on) — and for all we know it must be assumed that they will continue to utterly depend upon very roughly the kinds of ways of sustaining ourselves nutritionally etc. that we currently have.

Climate breakdown as 'a diffuse object'[42]

Consider a different kind of objection, this time on the grounds of human psychology: that there is a difference between immediate threats and 'diffuse' threats, and I am perhaps underestimating that difference.

For example: if you asked a group of parents whether they'd risk their lives to save their child from being hit by a car, most would say that they would react instinctively and risk everything for their kid. And they're probably right. However, our instincts (and that is what this sort of love typically amounts to) don't stretch so easily to very distant or diffuse threats — partly because, through virtually our entire evolutionary history, imminent and specific dangers have been the major risks to individuals and families. It's only in the last blink of an eye, in evolutionary terms, that distant, giant threats to our survival — nuclear weapons, man-made dangerous climate change — have come close to the forefront of our awareness. In a nutshell: our psychology may make us unreceptive to the argument I've presented, even if that argument is, in theory, compelling. We just are better at protecting our kids from speeding cars than at protecting our kids' descendants from the climate-dangerous greenhouse gases emitted by those cars.

There is some merit to this objection, and it partly explains how, as a civilisation, we have got ourselves into this awful mess. Human-generated dangerous climate change is a diffuse object; it is everywhere and nowhere, invisible as well as dramatically visible in its effects. It unfolds only across vast stretches of time. (Furthermore, its vast reach means that, uncannily, we can hardly talk about *natural* disasters any more.)

People associate love with the instinctual, focussing on reflexive, short-term acts of care for their children while long-term problems go unaddressed. Nebulous and diffuse threats like the climate crisis are harder to fight: their politics are thus inherently more difficult. But here's the crucial point: none of this affects the *logic* of my argument at all. It just makes it all the more important that it be grasped; more important that this rational-emotive case I'm making is understood, and then personally and politically acted upon. I am arguing that a parent would not really be as loving to their child as they thought if they did not extend their care to their temporally, geographically distant descendants. Because, in order to really care for their child, a parent needs to help facilitate adequate care for their children's children, and in order to really care for *those* children, they have to manifest care in turn for the great-grandchildren, and... You see where I am going with this.

And the 'they' here is you.

You can't provide this care just by dashing out in front of cars, so to speak. You can't even do it by making your children very rich: it would be rash (especially since 2008) to assume that our present banking systems will be eternally robust, rash to assume that an extremely unequal society will remain a stable one, and especially rash to assume that, in the very long term, any riches assembled for a large extended family can be maintained in the face of massive global ecological threats, such as escalating climate chaos or systemic breakdowns in the food supply. Given the gravity of the eco-systemic — and climate — collapse we are facing, everyone — rich or poor — will be exposed to these risks sooner or later. The super-rich might well be able to prolong their existence and that of their progeny for a decade or three in a fortified hideaway in New Zealand. But if the world tips into runaway climate change of the kind that Chapter 1 indicated is, tragically, now entirely possible — if, for example, the methane dragon is unleashed[43] — then decades may be *all* the extra time that even an inordinate amount of money will buy you. And that's no good if, as I have suggested, what matters to you is truly long-term regeneration across the coming generations.

Thus, the super-rich can only truly love their own children by putting their wealth at the disposal of the common good, to save the common future. Think long-term enough, and your descendants meld into a worldwide community, as I have set out above; any approach that is not *truly* long-term, across many generations, is simply not a viable way of tending to your children.

In sum, you can't save your children's children's future from the white swan of climate catastrophe by dashing out in front of cars, nor even by trying to make them very rich. But you just might be able to do it, as I shall discuss in subsequent chapters, by (for example) staunchly committing yourself, alongside others, to success for a revitalised global green movement.

It's a question of developing ways of caring and acting upon that care that go beyond our instincts. Because, unless we do so, we *aren't* actually caring as much as we most dearly want to for those for whom our care is instinctual.

The 'common-sense' view privileges people we know and have meaningful personal relationships with over distant descendants we will never meet. In response, we can now say this: the 'common-sense' view, while tuning into a psychological limitation that most of us have, turns out to manifest an insufficiently deep looking, a lack of love in its true sense. Yes, there may appear to be real obstacles in the way of *loving* our distant descendants: can one *love* the (as yet) faceless and formless? But the point, the great need, is to overcome those obstacles. And after all, was true love ever easy? Love looks on tempests, the storms of our grandchildren, and is never shaken.[44] Love conquers all — or it's nothing much if it has nothing to conquer.

And if it is the word 'love' that is causing you trouble in acquiescing to my line of thought — if it seems impossible to *love* the distant and formless — then that's easily solved; just switch to its near-synonym that I've already referenced extensively: care. The care you lavish on your children: it is that which has to shine onward into the further future if it is to be real.

Picturing the thread of time

Is this still about climate?

If you are still finding the discussion so far too abstract and are more comfortable sticking with your instincts, then let me address those directly. I want you to do something for me.[45] Close your eyes. (Well, close them after you have read the rest of the instructions...) Close your eyes and picture a child of yours — or your niece or nephew; or any young person close to you — at age five. Picture them enjoying their fifth birthday party. Take a moment, seeing them blow the candles out on the cake. Now fast-forward to their fiftieth birthday. Get a sense of who they have become. Their face now, what matters to them in their life, what they find most meaningful. Again, watch them blow the candles out. And fast-forward one more time. Now it's their ninety-fifth birthday party. The world, however difficult it has become, has become such that it is possible for them to live to ninety-five. Imagine this person blowing out the candles on their birthday cake one last time — and then being called upon by the children, grandchildren and great-grandchildren who surround them to make a valedictory speech. And as they begin their speech, what should pop into their imagination but you. Your face. So they speak about you: what you meant to them while you were alive, what you did that mattered, what your life's meaning and power was.

When their speech has ended, open your eyes again and write down their speech about you. The speech you want them to be able to have given.

That's the end of the exercise. (Don't just read about it. Do it. It's worth ten minutes of your life.) Try to do it without getting egotistical about it. It's not really about *you*. It's about taking seriously the passing down the generations of the splendid bright-burning torch called life. And it's about therefore starting to put yourself in right relation to all those descendants of yours who will be gathered in that room for your daughter's/niece's ninety-fifth birthday party. In other words: it's about ensuring, to the best of your ability, that the glorious chain-link of love that binds you to them and theirs is able to flourish, and that the scene you have just imagined can really happen... That you really will have great-grandchildren, and that love can go on living.

Life is this astonishing thread reaching from the deepest past, on beyond your death, into the furthest future. You are an entirely integral part of this thread. And they're right there, your descendants, closer than close. They're real — except, unlike your ancestors, your descendants haven't arrived yet. That is why they are infinitely vulnerable and need infinite protection.

The forward-looking loving care we manifest for our more distant descendants — called for by the circumstances of our world and of our species' power over the very future — needs to be equal to that that we manifest for our kids. Or rather: the love we manifest for our kids becomes *equivalent* to the care we show the deep future — and vice versa. For our distant descendants are what becomes of us. Given that we care for our kids, we need to find a way to care (equally) for their kids, and there is no end to this care and this connection.

The kind of imaginative exercise that you have just done could and should be repeated by every generation, generating and regenerating love. It threads on into the future, spinning kindness off at every iteration.

It's inconsistent, it's unloving, not to act accordingly. The implications may well be difficult, challenging; we are called by our own care, once we think it through, to rise to that challenge. To commit to meeting it. And after all, is the challenge really that hard? Especially after one has carried out an exercise like that I offered just now. Aren't I simply saying something that, at some deep level, we all know: that nothing can be more important than building a secure tomorrow; that, if we were to make our kids happy for a while at the expense of their kids, then we could hardly be said to have cared for our own properly? Isn't the logic that I am drawing attention to, that we claim to find psychologically difficult, actually a restatement of what is and ought to be common-sense?

You can no longer build a secure tomorrow by scrimping and saving so that your kids have a better life than you. For them to have a secure life at all now, something very different, much more collective and larger, is going to be needed.

The essence of our lives

Let me consider one final objection: that there *is* a difference between caring about someone and caring about what they care about. Does that point undermine what I've been saying? Much of my argument in this chapter rests on the idea that if you care about someone then you care about what they care about. On reflection, does that necessarily follow? I may love my eccentric aunty, but I don't care much about lots of things she cares about — and, in fact, one of the ways I care about her is wishing that she would concern herself with some of those eccentric things rather less...

I sympathise with the sentiment of this objection. But the answer to it was already outlined, earlier. It is this: you wouldn't really be caring for your aunty if you frustrated the *essential life-project* that gave meaning to her existence — especially given that, in the case of your children, we have established that you share the very same essential life-project (i.e. caring for the next generation). This follows from my fundamental starting-point: that it is reasonable to assume that what one values fundamentally will also be what one's children value. I, therefore, have a riposte to the possible objection that just because our grandchildren love their children doesn't mean that we should or do love those great-grandchildren. What I have just shown is that that claim is irrelevant because it gets things back to front. What matters is that one can only properly love and care for one's children *by* extending to one's grandchildren the opportunity to flourish.

And this is what matters: it is what ensures human lives' meaning.

One cannot have a decent life without gifting one's kids with a decent life. This is not negotiable. And destroying the planet's future capacity to support long, happy and healthy lives, with all the incalculably vast suffering that this would entail, frustrates our children's essential life-project. It thwarts them in the very project — that of caring for one's kids — that is manifest and urgent in our care for them. And this brings out something rather beautiful about the line of thought I'm essaying in the present chapter: it is exactly that care which gets passed down from generation to generation — deliberately, implicitly, and willy-nilly — and ensures precisely the continuance of the generations. *Obvious!*

It is this surety of the future that our species' breaching of planetary limits is putting into question, as outlined in Chapter 1. How could we justify depriving our kids and their kids of the very thing that we non-negotiably desire and care about: the possibility of bringing children into the world whom we can love and protect?

Essential to our very existence as human beings is our role of passing on the project of caring for the future to the next generation. Without such care, we are no longer ourselves. And this central project of ours, I argue, necessarily entails the essential duty of caring for the deep future. The essence of our lives is our love, and our love iterates or else it isn't real.

All of that having been said, perhaps you (still) think that at the end of the day the love you feel for your children just cannot be equated to what you feel for your grandchildren (or if for your grandchildren then not for their children), or that at some point in the chain-connection of love there will be, there *must* be, a diminishment in your feelings. If you conclude, then, that your grandkids just don't matter so much to you as your kids, I would ask you this: how you would feel saying that to your kids (or your grandkids)? (Especially once your kids point out to you the existential peril that your grandkids are in.) I think your children might say things back to you like, 'That's lovely that you love me so, but there's something terribly missing if you don't extend that love to my kids too. Don't you see; they matter so to me — you must know that — so I don't *get* it if you won't look to them and help look after them just as passionately too. If you love me, Mum/Dad, then you love them too!' And my point is this: the same emotional logic plays out at *every* point in the chain, the great unfolding chain of human being, which is also a chain of love.[46] Loving your kids, then, really does amount to loving your descendants to the nth generation, for any and every value of n.

What is to be done?: a test-case

A smart response to the line of thought I've set out in this chapter could go roughly thus: 'OK, I'm convinced that we need to care for the future; but the way we best do this is by taking care of our immediate successor-generation(s), and then allowing them to express the same care for their immediate successor-generation(s). We should take care of our kids and allow them to take care of their kids. That is how care best iterates down the generations!' There is something right about this. We ought, we *need*, to allow space for our descendants to care for theirs, and there's no alternative anyway to doing this: they will be closer to the situation, they will have more knowledge and agency and (one hopes) wisdom in it. A great way to express the core of our task then is this: we have to keep getting ourselves out of the way so that they can do this. We need to stop ourselves from doing anything that would stop them from having that chance, of caring for their own.

But here's the thing: even that task requires that we actively look far more than a generation or two ahead. Especially now. Because our power to stop or to mess up the future has become so great.

Chapter 4 of this book deals more fully with the practical implications of my argument for policy. However, let me give you an initial indication now of one such practical implication, to illustrate the point I'm making about the need to look and to protect, actively, far ahead. Projecting forward into the exorbitantly long-term future in the way that taking climate seriously demands is difficult, but the following example will help concretise my argument.

Let's think about nuclear power. It generates toxic flammable waste which remains deadly for hundreds of years, and which must be stored somewhere on our planet for thousands of future generations to come. If you care about your kids, and if you are only thinking of them and not the distant future, then you might take the attitude that nuclear power is good: once it is built, it will benefit your children by providing them with power — power that is allegedly low-carbon, at that. But once you are thinking just as much about your great-grandchildren, and indeed your many scattered great-great-great-great-great-great-great-great-great-great-great-great-great-great-great-grandchildren, and so on, then nuclear power starts to look a rather less attractive option. Because they get no power — only poison. This will not do. Given where this chapter has led us so far, we have to find a way of protecting our progeny — which is, it turns out, basically everyone — in the very long term. Any way forward which is not compatible with very-long-termism needs to be rejected. Thus, when we are looking for short-term fixes to provide ourselves with less climate-dangerous power sources than fossil fuels, we had better look very hard elsewhere, rather than to the nuclear option. From the perspective outlined in this chapter, searching for energy sources that are not going to poison our distant descendants looks like nothing more than an elementary precaution.

This is especially clear once one takes seriously the possibility of societal collapse.[47] If our civilisation disintegrates — which is very likely to occur within the lifetimes of some readers, unless we manage to effect the kind of extraordinary boldness and transformation I advocate for in the latter portion of this book — then the last thing our struggling children will need, and the last thing they will be equipped to deal with as they attempt to survive in the vestiges, is out-of-control nuclear fires across the globe. And that's what may happen unless nuclear waste is managed in cooling ponds and tended carefully for hundreds of years: it may burn, spewing poison across the globe, for decades or even centuries. Given that we are uncertain that we will be able to prop up civilisation through the coming unprecedented crisis, it would be utterly reckless to build more nuclear infrastructure, as many current governments (including the UK's) are presently doing.

The details of the specific example I've just given might be contested. Regardless, the broad point remains intact: that future safety cannot be assumed from present performance, across vast reaches of time including potentially the collapse of civilisations. What my example seeks to illustrate is a more general truth that I believe I have established in this chapter. What is *not* contestable is that the kinds of problems facing us now ask us to think much harder and on a global scale, and across long reaches of time, when we know that long reaches of time are relevant (as they are, for example, in the nuclear case). We must take seriously the foreseeable (very-)long-term consequences of our ways of life and political choices, especially when those choices may portend serious or irreversible harms.[48] If one thinks about one's kids in the serious and consistently applied way I've argued for here, one *might*, for example, still come to the conclusion (as a minuscule handful of ecological thinkers have done) that nuclear power remains a good option for humanity. But one could not — as too many are prone to doing now — simply decline to even consider the truly long-term ramifications of going nuclear. Nuclear waste is a great test-case; it shows clearly that it is not enough for each generation to plan only a generation or two ahead (nor even seven generations). Nuclear waste is like an arrow that we fire into the far future[49]: we don't know where it will land or who it will kill when it does. One couldn't fire an arrow towards a distant crowd and 'defend' one's action by declaring airily that, because one has no idea who exactly where the arrow has landed and so who one may have killed, one hasn't done anything wrong. We can't know exactly who our descendants will be, but that doesn't alter our responsibility toward them one jot. In fact, their utter powerlessness in the face of our actions should accentuate our sense of care toward them, much as the powerlessness of a baby is precisely what tugs on us to take care of them. People think that because we don't know all the effects of human-triggered dangerous climate change for sure, we don't know (for sure) that they will definitely be very bad, so it's OK to risk it. Once again, such uncertainties do not count as exculpation.

If one cares for distant generations, then one will tackle threats to them; and if we have voluntarily set in train processes or created materials we shouldn't have, then we have betrayed the future.

I have argued that one *does* care for distant generations, by virtue already of caring just for our next generation. As humans, being primates (mammals), inexorably and beautifully do.

And so one is bound to draw the consequences.

We ancestors, we mammals

This chapter as a whole argues that caring for your kids equals caring for the very deep future of humanity. With the nuclear example, I reminded you of how there are some problems that reach into the very deep future (dangerous climate change itself is another such problem: we have already probably set in motion the eventual melting of the ice-caps, and this will have dire consequences after we are dead and buried). Nuclear power/waste is, therefore, a clear example of something we should be dealing with differently by taking into account the long-term wellbeing of our descendants and taking *care*.

What are we? We are creatures who care for our children. We are defined by this care; especially as human children are defenceless for longer than any other kind of young, longer even than marsupials (which, when they are born, and head to the pouch, are just a centimetre or two in length).

We are literally nothing without this care.

So, what we become in our essence is ancestors.[50] Now let me suggest something that might sound counter-intuitive. From the perspective of being an ancestor, there's a sense in which you are already gone, already nothing; for you are quintessentially succeeded by your descendants.

But, of course, there is another sense in which you have agency: the very agency that your descendants need you to have.

Only if you embrace the former sense can you truly embrace the latter. You will not truly care for your offspring if your existence is focussed upon itself, yourself. You will only truly care for them if you manage to accept what is hard to face, what all of philosophy could be thought of a training in facing up to: your own mortality. Only if you really accept that you are here to no longer be here will you *live* now, wildly, passionately, deeply — and for the beyond-you.

The same is true on a societal level. We will not truly face up to the dire likelihood of eco-driven societal collapse, unless we already think of ourselves collectively as being gone, our civilisation as already being finished. Paradoxical though it may sound, we will only find the agency to mitigate this collapse if we are entirely serious about its being enroute to happening.[51] If we keep pretending that it is only one possibility among others, then we'll fool ourselves that it can be avoided without a complete *bouleverse ´* of our society and ourselves.

We have had an example of this lately. A planetary brush with mortality. Covid-19 saw society moving mountains to look after the old. Well, now is the time for intergenerational solidarity. As we emerge from being dominated by the coronavirus crisis, it is time to decisively turn our efforts to the ecological crisis and the untenable ways of living that largely caused it,[52] which can either be worsened or eased depending on how we choose to emerge from the Covid-19 emergency. We protected the old from coronavirus (or at least most of us genuinely tried to); we need now to protect the young from climate nemesis. And that process of caring really does start at home.

If you really care for your kids you will care *just as much* for what they deeply care for — their kids — and so on, indefinitely into the future. If you don't care as much about your grandchildren as you do about your children, then that is, it turns out, just a roundabout way of saying that you don't care about your children as much as you claimed to. (Directing my attention to the older reader, for a moment: be grateful, dear older reader... because I've just given you a reason for feeling OK about the fact that, more often than you might like to admit, you maybe feel slightly fonder of your grandkids than of your kids...)

The beauty of this argument is that, once you've understood it, you see that it counters the very parochialism that could seem inherent in caring just for your own kids. We need the rich countries of the Global North to take ownership of the wicked problem that is climate, but the effects of the climate crisis are currently mainly felt in the Global South. My line of thinking gives us, at last, a way of threading these together. For the sensible thing for you to assume is that your descendants will not be restricted to the country you are from but will be global. Thus, the only sensible way to parent our future is to provide the Global South with exactly the help it needs. Caring about your own kids *implies* serious seeking to achieve global justice.

Whichever way civilisation crumbles or changes, love for our kids has to last forever, down the generations. We started this chapter with a small, uncontroversial premise that turns out to lead us to a dramatic, powerful conclusion. An emotionally powerful and exciting conclusion because a conclusion that will have consequences. With this lever, we really could move the world. In a global situation that can appear hopeless, perhaps really understanding our mammalian nature — in the way that I have sought with you to re-understand it in this chapter — can give us new hope.

Our children and (still more so) their children (and so on) are profoundly dependent upon us. If we fail them, at this fateful historical juncture, they will have no recourse. Our failure will be fatal. We *have* to try, for them. In this chapter, I have equipped you with some new tools for thinking about how that effort can take shape. Everyone cares at least for their own kids. But that means that each couple ends up being parents, guardians, of the entire deep future. We can raise our gaze, out of the short-term greed which our consumerist culture has dragooned us into, toward the far horizon of time. That is the real meaning of your love for your children: a care for the whole future of humanity, so far as we can see, and so far as we need to see. *That* future is what, together, we have to try to *real*-ise. To make real. To safeguard. Indeed, to treasure.

We do that best by doing enough to ensure that we do not blot out the far future. That our kids have a chance to express the same heartfelt care for their kids and their distant descendants alike that we do.

Let me give the last word in this chapter to a mythic voice that expresses just this quite splendidly. In the last book of *The Lord of the Rings* trilogy (1955), in what is called there 'The Last Debate', as the main characters try to come up with some desperate stratagem that will give them even a sliver of a chance of the future not becoming darkness visible, Gandalf says this: 'It is not our part to master all the tides of the world, but to do what is in us for the succour of those years wherein we are set, uprooting the evil in the fields that we know, so that those who live after may have clean earth to till. What weather they shall have is not ours to rule.'[53] Yes, we look to leave behind clean earth for the generations who will live after we are gone to be able to till. What's changed though, of course, since Tolkien wrote these words, is that our industrial hubris threatens not just to wreck as it were the forests around Isengard and ultimately the Shire itself, but the very weather. Our task is not, overweeningly, to rule the weather; but it is to mean it when we say that we are going to seek to roll back the *wrecking* of the weather, of the climate, that our self-cancelling civilisation has begun. That is a task a little like building a cathedral, only immeasurably more important and vaster. It is a task of *building down* our excessive impacts upon our common home. It is a task for now, and for the ages.

So far, this book has laid out a logical progression demonstrating that what you feel in your heart for your own kids in fact equates to a deep care for your descendants, even those that you'll certainly never meet. Now let's turn to asking what the consequences of this are. If, by virtue of caring for my kids, it turns out that I also care for the whole human race, across its entire future, well — so what? What follows from this? What, in particular, does it mean for my attitude toward the rest of life: toward other animals, toward the Earth? The next chapter shows how concern for the whole human race in fact also entails caring for the planet. If you really love your kids, you owe them and theirs a planet inhabitable and good, far into the future. (From there, the book sets out in later chapters to answer the most urgent question: what then is to be *done*?)

3

Truly caring about the human future means truly caring for the whole Earth

'You and the land are one.'
– *Excalibur* (1981)[54]

From placing humans centrally to placing nature centrally

In Chapter 2, we established that real care for your own children logically equates to real care for the deep human future. Let's assume you now grant that. Let's say that you see that you care about human beings across the world and distant in time from us. But, using the same procedure as I did in Chapter 2, I'm going to treat this as a minimal premise that doesn't demand anything more than what we've clearly entitled ourselves to. Let's assume, that is, that you only care about that; that you don't care about the non-human deep future. Let's assume, conservatively, that you place human beings at the centre of your worldview, and non-human beings enter into the picture only insofar as they are instrumental in assuring human wellbeing. This attitude is known as 'anthropocentrism'. It's a simple, widespread, seductive, and understandable notion. It is frequently taken for granted in our world that human beings are at the centre of human concern, and that non-human beings only matter insofar as they help us. Let's assume that that's what you now grant, and no more than that. And let's see what follows.

In other words: I'm not going to do what some 'green' thinkers do at this point. I'm not going to lecture you and say that you ought not to place humans so centrally. I'm not going to browbeat you into feeling bad for not caring more about the more-than-human. Instead, I'm going to see what follows from a genuine effort to place humans at the centre of one's concern. I'm not going to assume that what follows must be bad. I'm going to see if it might be good.

For remember that the previous chapter established that this anthropocentrism is not just a short-term caving into present human desires. An egocentrism writ large. Far from it. It needs to be properly long-termist. *Very* long-termist. One is hardly placing human beings — including those many unknown generations to come, and who your kids are destined to become — at the centre of one's concerns, if one allows things to be done now that massively negatively impact future generations. True anthropocentrism needs to be smart, and prudent. It needs, that is, to be *long-sighted*.

So now here is the argument of the present chapter in a nutshell. A truly prudent anthropocentrism equates to ecocentrism: placing nature as a whole, the entire planetary ecosystem, at the centre of our concern.[55] For one cannot take care of humanity without caring deeply about increasing our resilience, including crucially through maintaining the biodiversity — the web of life — on which we ultimately depend thoroughgoingly. To safeguard our descendants there is no alternative to acting with genuine care and caution so as to preserve all this. These goals call us to protect the integrity of our ecosystems. In fact, they call on us to realise the extent to which we are indissoluble from, inseparable from, the ecosystem in which we are nested. In short: we are nothing without a healthy planetary ecosystem — and the safest long-term way to ensure such health is to maximise the integrity of that ecosystem.

Care-ful thinking

Recall that, as things stand, we are losing biodiversity, we are losing life, at a heart-rending and terrifying rate: quite possibly a species every quarter of an hour or worse. And we have lost even faster the *volume* of wildlife on Earth: I'm in my fifties, and during my lifetime we have lost approximately one per cent of the wild living beings on Earth each year.[56] That's over half gone since I was born. (In the back-story of *Avatar* is the same kind of destructivity that is embedded in our present, and possibly worse still, our future. As the protagonist of that film says, contemplating the difference between the Earth he left behind and the new hope he finds in the world of Pandora, 'There ain't no green there [on Earth]. They killed their mother.')

Worse still, the projected rate of extinctions is scheduled to increase exponentially in coming years.[57]

This is a horror story. But for those unmoved by the horror, let me point out something else. It's stupid. Reckless.

The case I am making here — the logical argument, based on minimal assumptions — can be simply put in terms of the Precautionary Principle that we started to encounter in the previous chapter. Just as we saw, in Chapter 2, that it would be reckless not to assume that your descendants will populate the Earth, this chapter demonstrates that it would be reckless not to assume that those descendants will most probably depend utterly upon the other — non-human — populations of the Earth. That is, the precautious attitude to take is to assume that, without a robust and biodiverse ecosystem, our descendants will find life on Earth at best very difficult indeed. More likely, impossible.

That assumption could conceivably turn out to be false[58]: we (our descendants) might conceivably turn out to 'escape' bodily limits altogether and become (for example) some kind of computer-upload (Perish the thought!). But it would be an absurd gamble to argue that this remote theoretical possibility makes it acceptable to let other species go extinct — just as it would be absurd to argue that, because bionic limbs might become commonplace by the time I'm old, there's no harm in doing something that risks having my arm hacked off today.

The more species and variations within species there are, the more resilient are our ecosystems.[59] Biodiversity is directly correlated with ecosystemic health. Thus, a *maximally* biodiverse Earth is maximally conducive to human health and resilience. There is not a cigarette paper's width between ecologically sane anthropocentrism — anthropocentrism that has truly taken on board the Precautionary Principle — and an 'ecocentric' approach placing our planetary ecology at the centre of our concerns.

Everything we know about the natural world suggests that protecting the ecosystem is vital for safeguarding human life — but this knowledge is, in many ways, still desperately thin. As I've laid out already in prior chapters, this lack of information is all the more reason to exercise caution. We have no clear idea how many species there are on the Earth,[60] for example: estimates vary from about two million to about ten million, an astonishing range of uncertainty. There is much we do not know about biology, particularly ecology. We have little idea what deep and unexpected mutual dependencies there are between different living (and non-living) beings. We are learning that genetics is far more complex than we had thought, in that what it takes for genes to express the 'information' that they encode is turning out to be far more open and far less predetermined than had been thought a generation ago. Much of the functioning of the brain is a deep mystery to us. It may surprise you to learn that we are only really now starting to understand what soil is, and that it is far more complex than even farmers had realised (at least in the modern-day West). Arguably, we don't really know what life is, or how — as the so-called 'Gaia' hypothesis contends[61] — it seems to have become able to regulate its own conditions for existence. The Earth system has become relatively stable for life until very drastically disturbed, but this fact is hard to compute, on the basis of biology. We know more about the surface of the moon than about the deepest trenches of our oceans; we know that there are vast hosts of life that are unknown to us; we know more about the Andromeda galaxy than we do about most extremophiles (creatures that live in extreme environments in our planet that are highly hostile to mammalian life).

On one hand, it strikes me that there is something rather beautiful about this ignorance — about the fact that, despite the profound efforts humans have invested in understanding our Earth's natural history over many millennia, so much of nature remains to us a compelling mystery. On the other hand, this ignorance can produce some dangers; for it implies that we might right now be harming parts of the Earth system that are far more critical to life (or at least to our life) than we realise.

The massive uncertainties that exist in this area yield the strongest argument that there is (starting from the kind of conservative, unobjectionable premises which I've restricted myself to in this book[62]) in favour of preserving all the nature that we possibly can. An evidence-based justification for the preservation of life on Earth in its details is not (yet) forthcoming; we typically cannot prove that any given species is indispensable. But our very ignorance is the best reason there is for careful thinking and cautious action, i.e. precautionary preservation: we just don't know what small, seemingly insignificant, species or ecosystem might turn out to be crucial to the whole: a 'keystone'. Indispensable! We *do* know that right now we are extinguishing many species that we don't even know exist yet; that species unknown to science are ending forever, right now, in our rainforests and elsewhere across the world. It isn't safe to gamble on the rest of life being able to do without them. Our profound ignorance, *still*, of how so much of life works makes for a powerful precautionary argument. The precautious thing to do, to make ourselves and our descendants safe, is always to presume against the destruction of unique ecosystems and species. And moreover, to seek to restore those that have been lost or damaged.

To generalise the point still further: our uncertainties about the Earth — far from meaning that it makes sense to carry on as we are pending further information — in fact mean that we actually need to change course now, to stop damaging the Earth in ways we are not yet even aware of. The more uncertain we are about how things work, the less we should allow our activities to interfere with them.[63] The Precautionary Principle, not to put too a fine a point on it, warns us not to mess around with things we don't understand.

Consider this principle in terms of the climate crisis. 'Anthropocentric' positions are obviously sorely lacking if they do not also accommodate and seek to prevent harms to ecosystems from dangerous anthropogenic climate change. But those harms, despite the tremendous accomplishments of climate science, remain uncertain in many important respects. Crucially, one such contingency is that we don't know just how vulnerable (or resilient) life will be against the unprecedented, dangerous climatic perturbation that we have released. This system-deep uncertainty[64] — one of many unanswered questions in climate science (others include the sensitivity of the climate to CO_2 increase, tipping points and feedbacks[65]) — does not, as 'climate sceptics'[66] stupidly suppose, make it reasonable to *assume* that the threat is not quite so grave. Rather, it points in the direction of our having to be more careful. The downsides of human-inflicted damage to our climate are incalculably bad; we simply have no idea just how severe they could get. There is an asymmetry here: if we turn out lucky, then we might even experience some marginal gains from the changes in our climate, but if we turn out unlucky — as looks increasingly likely — billions could die.[67] The downside massively, completely outweighs the upside. Uncertainty is not a reason for prevarication; it's a reason for strong precautious action. The less we know about the exact effects of our activities, the stronger the case for changing course and ceasing to mess with the climate.

If we take a precautionary approach towards the climate crisis, then long-term care for humans translates directly to long-term care for the Earth. At this moment in history, when we have exceeded various dangerous parameters (including climatic and biodiversity limits),[68] and may have crossed others that we do not yet know about — and so are no longer in a planetary safe zone — it is clear that caring about humans in the present day *means* ecological sensitivity, now and deep into the future. In other words, it has become staringly obvious that a properly prudential, ecologically sensitive anthropocentrism coincides in practice with a biodiversity-preserving, ecosystem-restoring ecocentrism.

Whatever it is that is required for long-term environmental flourishing, for a biologically rich environment, is also required for true human flourishing. We cannot flourish in a genetically impoverished world, or one without a wild — or, of course, in one whose future generations are condemned to decline. More immediately, we cannot flourish if our destruction of habitats continues to unleash a series of new pandemics caught from the disturbed and mistreated animals seeking sanctuary from our bulldozers. We cannot flourish if we lose the biodiversity that includes our pollinators, our soil fertility, our actual rich crop varieties... not to mention providing us with potential cures for diseases, or with potential new varieties and entirely new foods to us (and who knows how our need for foodstuffs will evolve in the coming millennia?). We would not be flourishing if everything were domesticated and nothing was a mystery, and there were no margin left for the Earth systems not 'in our control'. For part of what makes these natural systems robust is precisely that they take care of themselves, without relying on human 'stewardship'. (And this so-called 'control' is, anyway, far less reliable than it might seem. The more we perturb nature, the *less* reliable our control over our environments. This can be seen starkly in the worsening of 'natural' (i.e. mostly human-influenced, tragically) disasters in recent years — including in the emergence of Covid-19.[69]) In toto: we cannot flourish without the flourishing web of life, of which we are just one part.

In this era of climate emergency, guaranteeing human flourishing requires us, out of a prudent *abundance* of caution, to protect what we misleadingly term 'the environment'. It would be mad to do less. There is no way — in the real world, taking the long view, the safe view — of untying the bond between ourselves and our 'environment'. Because, we can now see, there is *no genuine dividing line* between the two once one takes serious account of precaution and of our distant descendants. In the face of the vast challenges of anthropogenic dangerous climate change and the human-driven degradation of habitats, biodiversity-preservation — primarily by way of ecosystems preservation and restoration — is a central way in which we can look after the deep human future.

And it's worth noting now a way in which the vitalness of habitats and the wrongness of their destruction is an even more credible place from which to begin one's considerations than is the wrecking of our common climate. For habitat destruction is utterly undeniable. The thin veneer of possibility for denial which climate-deniers have exploited ruthlessly is not present, when football-fields worth of irreplaceable ecosystems are going up in smoke. We should probably highlight this more, and resist getting dragged into residual arguments about exactly how bad climate breakdown is. For climate breakdown is one part of ecological breakdown — and the latter, when it is manifested in the simple form that has done most damage to date, of eliminating natural habitats — can be seen with our very eyes.[70]

There is no 'us' versus 'our environment'; there is only one seamless, beautiful web of life.

As my teacher Joanna Macy puts it: if we think ourselves, the *anthropos*, properly, we will eventually understand that we are not distinct from the sustaining environment at all.[71] We will then find that being 'selfish' — in the sense of taking ourselves seriously as worth preserving and continuing — in fact requires us to identify with the entire world and its entire future... As I put it earlier: the sane attitude to take is to see that we are (and should assume that we will remain) *nothing* without maximally robust nature to support and *co-constitute* us. We could think of the world not as something 'out there' that we love or need, but simply as our greater body. The ultimate resting-point of a wise anthropocentrism is a wise ecocentrism, one that sees ourselves when it sees the Earth and sees the Earth when it sees ourselves. There is no keeping the human and the non-human separate. We belong to ecological communities, which also constitute our identities: this is what it is to be biological beings. This is why indigenous peoples have often said that we belong to the Earth, rather than it belonging to us...

Open your eyes and look in the mirror[72]

I hope you understand what I've been seeking to say, that it makes sense to you; perhaps alters your perspective or moves you. But if what I've been saying here is a little too abstract for you, let's see if another imaginative exercise can bring the idea to life.

This exercise requires you to spend a moment of your time outdoors, and with someone else, preferably somewhere not too public. Once you've found such a spot, take it in turns to be lightly blindfolded. And then one of you leads the other around by the hand, slowly. The blindfolded one may find themselves able to notice sounds or smells that they don't normally notice. The guide should invite the hand of the guided one periodically to touch something, anything. Bare soil, the trunk of a tree, a daffodil; anything. (The first time I ever did this exercise, I was delighted to be able to guide the hand of the person I was leading around onto a cat, which they then got to briefly stroke.)

And then once in a while, lower your partner's blindfold, direct their gaze, and give them the invitation: 'Open your eyes and look into the mirror.' And the 'mirror' might be: ... anything. A bush; a bird on a table; some grass; the surface of a pond below them reflecting their image back; the sky; another person.

This exercise is called The Mirror Walk. It is an invitation to let go of one's sense of separateness from the natural world, a sense that our society unwisely encourages. Specifically, The Mirror Walk invites one into the experience of seeing 'others' — other living things, other parts of the ecosystem — as 'mirrors' of oneself. You are the grass; and the grass is you. What you see, whatever it is, reflects back your extended self, without which/whom you wouldn't exist.

Please do this exercise. There's no substitute for experiencing it first-hand. It has a beautiful power, I promise you.

Eco-activists' arguments, beginning from different, 'stronger' premises, might ask you to care for something that is seemingly 'other' to you. Persisting with the anthropocentric tradition (properly understood), as I have sought to do in this chapter, has significant advantages over this. In particular, starting from the anthropocentric tradition can lead the many human beings who still fantasise that ethics can be first and last a human affair down the winding road on which, together, we desperately need to travel. On which, in fact (and whether we realise it or not) we *long* to travel. Among philosophers, policy-makers, economists and ordinary citizens, the dominant view remains that human life is the alpha and omega. Only if we overcome this assumption do we have any chance of surviving the future. But what I have shown in this chapter, building on the last, is that we can most forcefully overcome it *from within*. By starting unapologetically from the anthropocentric tradition.

Retaining this tradition can lead us all to what I suggest in this chapter is the destiny of anthropocentric thinking: a care for the 'more-than-human', for ecosystems, that is profound and thorough-going. One that is serious about recognising our commonality with the rest of life. A *shared* existence. That is what ecology means. After all, there is no waste in nature, only circuits. To come back to the root meaning — we might call it the '*home*' meaning, even — of 'ecology': etymologically, it means home/dwelling. Our natural home. How could we possibly be estranged from this?

When we look at the world in this way, we can see glorious difference, but simultaneously we see something that's not estranged from us at all. The world mirrors back to the eye of good faith the existence of ourselves *via* the existence of fellow beings, human and otherwise. Our companions; not mere 'resources' or objects of domination.

For the love of life

We need to take a good long hard look at ourselves — and if we really do, with a wide enough gaze, what we see is indescribably beautiful. Viewing the world in the way that I've just set out makes it possible to start from a conservative position, placing ourselves at the centre of our worldview, and then vastly expand our sense of the 'non-human' world and its utter value. Its closer-than-closeness, and yet its difference.

The stubbornest 'anthropocentrist' can legitimately care about *whatever* members of the human species care about, for starters. And humans care deeply about all sorts of things: including their (our) ancestral lands, companion animals, wild places, big open 'empty' spaces, rare species, familiar species...

It was always a very crude and partial anthropocentrism, an anthropocentrism decidedly unflattering to human beings, that saw us as having no essential interests in other beings. Only an abominably weak or nihilistic vision of humanity could assume that the only things we care about are ourselves. We care about more, that is, than just humans. It does us a disservice to suppose that we do not (for instance) have a tendency toward reverence for nature and the wild, a responsiveness to other beings' suffering, and a care for *all* our kin. Such a supposition helps in suppressing these fine tendencies.

The true good sense available in the word 'anthropocentrism', I'm suggesting, involves centring yourself in *the human*, in all the depth and splendour of that concept (i.e.: of *us*). In a universalistic sense of ourselves, and a sense that turns out to extend way beyond our species. It is part of what it means to be anthropocentric, then, to care about all the diverse things that human beings care about, and to care about things (rather: beings) other than one's own direct, human kin. Even if you refuse to grant that we are inextricably tied up with the nonhuman world — that we are inseparable from it — the fact remains that one of the things that is so distinctive and fine about the *anthropos* is its care for what is superficially 'other than' itself. Anthropocentrism need not focus exclusively on humans. In fact, if one insists on placing humans at the centre of one's concern and tries to exclude everything else, what one has come up with is not so much anthropocentrism as 'anthroponarcissism'. Whereas the version of anthropocentrism built on here in this my essay is a celebration of humanity in its fullest expression.

This expansive sense of humans' capacities for attending to and caring about what is superficially separate from or 'other' to us has become literally vital. For we are going to need this all-embracing concern for the world around us if we want to survive what is coming for us and those we love.

We are at our best, we are most humane, when we stand up for those who are most voiceless — a group that includes, notably, non-human animals as well as our own descendants.

What I've been talking about can be put in terms of a splendid idea of the great biologist E.O. Wilson's.[73] His term *biophilia*, coined in a beautiful book of that name, suggests that life loves life; that, in particular, human beings are spontaneously drawn to life. Wilson argues for this claim on numerous plausible bases, including the increasingly well-documented role of nature in wellbeing (illustrated by the famous experiment showing that hospital patients recover better if they can see a *tree* from the window of their ward) and our aesthetic preferences (Even across different cultures, humans seem drawn to natural spaces and representations thereof, in ways Wilson lays out). So, while anthropocentrists traditionally want to claim that humans 'only' value humans directly, and non-humans instrumentally at best, healthy human beings are 'biophilic'; they care about the nonhuman more than is dreamt of in a narrowly anthropocentric philosophy.

Let's briefly review what real humans value. Human beings value their children. They (we) value the fact that there are things beyond price, that there are things that are unknown and places that are untouched by human hand; and (some) human beings also value, for example, fast cars. But it is implausible that they (we) *need* fast cars, or that the experience of valuing fast cars is a necessary component of a fulfilling human life. Moreover, fast cars (which depend on limited natural resources and emit deadly forms of pollution) are products, commodities, that by and large are not ultimately compatible with valuing our children, nor with valuing nature — which, I have shown in this chapter, is essential to caring for our descendants. Following Wilson, I'd assert that human needs are wider than has often been assumed: in order to flourish, human beings *need* certain things that cannot be valued like commodities. They (we) need green spaces, nature, and so on (and, as noted above, pretty powerful psychological evidence increasingly backs up the biophilic hypothesis). We love these things, and we can't flourish without them. Humans are biophilic: valuing and loving nature is vital for our wellbeing. Living a fully human life turns out to require properly valuing the non-human.

I'm not really trying to convince you of anything here; I am seeking to remind you of your biophilia. It is something you can feel for yourself, when you do The Mirror Walk, or if you simply go to a favourite place in the countryside.

This love for life is perhaps best understood as the ultimate expression of our love for our children. For what we've seen in this chapter is that care for our children *calls* us to care for the vast systems of living things we find ourselves entangled in; that without an all-embracing love of life, we can't fully succeed in caring for our children.

Our Archimedean point

Let's take stock. Let me summarise how to see my line of thinking in this chapter in the light of what was already achieved in Chapter 2. Caring for our kids requires of course that they have a decent world in which to flourish. No one exists in isolation, certainly not children. If the world we leave our children is incapable of supporting safe, healthy and happy lives, we will have failed in our duty of care. Many parents are painfully aware, in this age of rising and justified eco-anxiety, of the diminishing opportunities to save their children from growing up in a desolate world — from living lives that are nasty, brutish and probably short.

But to shy away from what power one does have to make the world adequate for one's children is not a morally acceptable way of responding to the problem. Thus (as we shall see in the remainder of this book) it is incumbent upon us to do what we can to contribute to the world that our children will grow up in. And this point too iterates onwards into the future, generation after generation after generation.

This chapter has dealt with the legacies of an anthropocentric worldview that goes right back to early monotheism and the dangerous notion that the world was provided for human beings. Anthropocentrists, who have dominated Western thought for a long, long time, place humans at the centre of the world, and have standardly suggested that we are the only beings with 'intrinsic value'; according to this view, things only ever have value because they have value for us, narrowly conceived. I've argued that we are better than that. Humanity is more than that. For it turns out that such narcissistic anthropocentrists haven't thought carefully or expansively enough what it means to speak of the human. They have over-simplified. I have shown in this chapter that — paradoxically — if one *really* places humans centrally, it turns out that one is, in extremely consequential ways, placing ecosystems (including, of course, non-human animals) centrally.

This idea was partially anticipated by the radical ecological writer Derrick Jensen, who has written:

> [W]hen you take a long-term perspective, the dissonance between anthropocentric and biocentric [ecocentric] viewpoints disappears, or at least becomes much less. I am excluding the perspective of those who eagerly look forward to a future ever more dominated (and ruined) by technology. Those who advocate a technologically controlled future are not only not taking a long-term perspective (peak oil,[74] anyone? How about overshoot[75] and crash?) but they're simply insane. They are not in touch with physical reality (that's what 'high technology' does — it separates us from physical reality). *They aren't truly even anthropocentric, but rather technocentric...*[76]

It's time to be at least truly anthropocentric. In other words, it's time to return to some old ways: and put life first, not machines. In the face of severe global threats (such as the dangerous human-triggered climate change outlined in Chapter 1), there is no way to protect your ever-widening family tree into the distant future by being very rich, building a fortress, or having lots of tech-power at your fingertips: you can only protect your many distant descendants adequately by helping to protect the entire world. Whether or not you have realised it before, if you truly care about your own kids, then you care about all your descendants just as much. As discussed in Chapter 2, the only sufficiently cautious approach then is to assume that your descendants will live all over the world and throughout the entire human future. And given that you can't know in advance which parts of the globe they won't inhabit, it would be irrational of you not to seek to protect it all. Furthermore, we have seen here in Chapter 3 that that protection needs to be extended way beyond the merely human world. It would be utterly reckless not to protect the entire ecosphere for the very long term. And so, of course, such protection must begin *now* (for extinction is, we must assume, irreversible).

Irrespective of what people think they care about (or don't), the very existence of everything they value is fundamentally dependent on maintaining functioning ecosystems. Once one sees that fast cars and fast fashion are incompatible with the paramount goal of safeguarding our descendants' futures, then the superficial appeal of standard narcissistic anthropocentrism is severely blunted. In this way, the present chapter comes to the remarkable conclusion that any sane/true form of anthropocentrism collapses into ecocentrism. This is because an ecologically sensitive anthropocentrism coincides in practice with ecocentrism — and no genuine long-termist anthropocentrism is any longer tenable that is not profoundly ecologically sensitive. Anthropocentrism and ecocentrism may appear superficially to value different things — but these things cannot be kept distinct. They are one. We and our world are one.

To say that you and the land are one may seem a radical assertion. But there is another perspective from which it is once more simply a new recasting of very old common sense. Don't we all already *know*, deep down, what I am saying here? Isn't it a wisdom that each of us is born into? Children intuitively sense a kinship with nature and animals. But somehow — at least, in societies like this one (things are very different in indigenous civilisations) — we train this out of them.

It is obvious to children that the rest of the world matters and that we are nothing without it.[77] It could only have come to seem otherwise through a drastic story of separation, a deadly dogma of dualism, that has somehow come to seem natural. We are haunted by a dualistic separation of ourselves from nature, possessed by a fantasy of 'having' possessions, of 'owning' Earth: what the great writer Ursula le Guin called the fantastic dogma of 'propertarianism'.[78] It is time to come back from this fantasy-realm, before it's too late, and to stand instead on common ground, in this still-exquisite material reality that is available to us.

Our Archimedean point is our standing here, on Earth. We can do no other. For there is nowhere else to stand, nowhere else to land. We make our stand here, for this, our beautiful and only home, and for our own, our kids. And making our stand for them, we now see, is making our stand for the entire ecosphere. Forever. And that means *starting now*.

In Chapter 2, we saw that caring for your own children calls into being a caring for the deep human future. And remember what we said in Chapter 2: you could hardly be said to care if you stomped on what your children, and their children, most value. And if your (and your society's) carbon footprint stomps, if you see beings or species as disposable, then it's time to change. Now, in Chapter 3, we see what the core of that change is: we see that caring for the deep human future calls for caring for the entire biospherical future, both in the deep future and now.

We need truly long-termist thinking, and we need emergency protection for the natural world *right now*. That is what it means to actually love our children as the foundation-stone, the fiercest expression, of our love for life.

If you've made it this far, you've done the bulk of the philosophical heavy lifting that this book demands of you. It's time to start harvesting the payoff. Chapter 4 will explore what the two twin needs just outlined require of our society.

There is something special about humanity in that we can come to think the very things that this chapter has been about. And in that we can deliberately *restore* biodiverse ecosystems and help life to repair; we can make amends and find balance.[79] We heal ourselves by healing the Earth. And we heal the Earth by healing ourselves...

But the best way we do so in the longer term is by effacing ourselves.

Finding a humbler place where we can't disturb the balance so much in the first place. We make ourselves safest by making ourselves less of a risk to our world... Chapter 4 will explore how we actually do that. How we ensure care for the future, for our kids and all that they and theirs will need. For keeps.

4

What is to be done?: immodest proposals to save the future

'Save our world!'
– Chant at Climate School Strikes across the world[80]

'Do we continue to nourish dreams of escaping, or do we start seeking a territory that we and our children can inhabit? // Either we deny the existence of the problem, or else *we look for a place to land*. From now on, this is what divides us all, much more than our positions on the right or the left side of the political spectrum.'
– Bruno Latour[81]

A vital temporal paradox

There is a paradox that expresses our condition at this desperate moment in Earth's history. We are living in a world consumed by short-termism at the very moment when we most desperately need long-termism. We have a desperate, urgent need... to recover the long, slow view. There's nothing more important right now than... pausing, and carefully reassessing everything. Doing so right, to last. Ensuring that we don't leap from frying-pan to fire. For example, our response to the corona crisis must not be to worsen the fires of global over-heat, as a resumption of rampant economic growthism would do.

And yet long-termism alone is not enough, either. For the paradox is doubled: *the long-term situation is so desperate that we need an immediate crisis response*. We need to treat the climate crisis as the emergency it is — without hurrying into knee-jerk reactions. It would be catastrophic if our emergency response were to undermine the prospects for long-term human flourishing on Earth. It may sound like an outlandish joke, but technologists have recently issued outlandish proposals to place mirrors in the upper atmosphere to deflect the Sun's rays and keep Earth cool. No. That would be insanely reckless. Rather, our emergency response, while treating something slow-boiling as genuinely urgent, needs to be compatible with, and usher in, a new culture of true long-termism. Eco activists have long been urging us, rightly, to 'Think global, act local'.[82] Perhaps 'Think long-term, act now' expresses the new paradigm we need.

This chapter explores how we can achieve those seemingly paradoxical goals. The challenge is to mount an effective emergency response without going from frying pan to fire. We can, I argue, act swiftly and decisively in response to the crisis while also fostering new ways of thinking and being that can lead us to a longer term harmony with our planetary home. This kind of relation to Earth and its ecosystems is already being practiced. For example, the Iroquois Confederacy, a collective of indigenous groups in North America has thought of the next seven generations when making big decisions for centuries.[83] How many times have you heard a politician say, 'Let's consider what the effect of this decision will be 150 years down the line'? If only.

Or to take an instance from our own history, consider 'cathedral thinking'[84]: the remarkable kind of forward planning implicit in deciding to build a structure that you know will not even be finished when you die. We once again need the kind of thinking that looked far beyond the horizon of a single human lifetime: the kind of thinking that seems to have come naturally to people in medieval times, but is so hard for we 'sophisticated' ones...

The first part of this chapter sets out the single institutional change that, in my view, will be most conducive to moving our society to an emergency footing for dealing with the immediate crisis of the next five to ten years or so. The second part discusses how the emergency measures I propose could lay the groundwork for a more permanent institutionalisation of ecological long-termism at the highest levels of government. Finally, I focus upon the principle that brings together the prevention of ruinous emergencies and the creation of a route to safety for the deep future: precaution. I set out succinctly how and why the Precautionary Principle should be implemented in both short- and long-term policy deliberations. This phase of my argument articulates how this seemingly unassuming idea can drive the major shift in paradigm and vision we so desperately need. Toward a long, slow consciousness of the urgent and the emergent.

Pulling together in this emergency: the case for Citizens' Assemblies

This book has so far been a deeply impassioned yet logical exercise in truth-telling about the desperateness of our shared plight. Telling the truth about the extremity of the situation, and about the scale of the switch in thinking that is needful — if your children and all that they will become are to be made safe — is a prerequisite for being willing to *act immediately* on an emergency basis so as to sufficiently ameliorate it.

These just happen to have been Extinction Rebellion's first two aims: telling the truth about the emergency, and then acting with sufficient speed and resolve to deal with it. The latter, as was implicit in the previous chapter, entails answering with our society's actions the threats we ourselves have created to our climatic stability and to our ecology more generally. And so, we come to the crucial third aim of XR: to revive or create a real democracy that can envision a path towards achieving the second aim, that of acting sufficiently and in time. A real democracy could bring about wide and informed assent for that path, and there will be no real societal solution to the ecological crisis without such acceptance. In the absence of broad democratic consent, the kind of society-wide mobilisation that is going to be needed to head off ecological collapse just can't happen. Think of the way that China is so afraid of its own populace that it feels obliged to pursue rapid economic 'growth' despite the disastrous ecological consequences. Even the world's most powerful dictatorship is in no position to create the 'ecological civilisation' that it talks about[85] — unless it can get its citizens on board with a different, saner path (as many of them increasingly desire, to reduce chronic air pollution, re-beautify the land, not to mention making new pandemics less likely). If you are sceptical that the Chinese Communist Party is scared of its own citizens, consider that it banned the film *Avatar* from cinemas, for fear that the film would ignite mass revolts against land-grabs.[86]

Dictatorship cannot bring us to ecological safety. But nor can the so-called 'democracies' that we currently have. With the entirely patent failure of representative democracy to head off or even cope with the ecological crisis — in fact, countries such as the UK have been responsible to a large extent, including through colonialism and its legacy, for *creating* the crisis — it is time to let the people, presented with the stark truth, decide how to prevent nemesis. How to change everything so as to face the emergency and force sufficient right action. To achieve the drastic amelioration of this mother of all crises that is Earth's most desperate need.

The first demand of Extinction Rebellion is to describe and accept the situation: not an easy task, when the situation is as grim as it is. Just facing up to reality at this point demands a forceful act of will. The second demand sets out the kinds of targets we need to achieve if we are to avert catastrophe: the as-swift-as-possible reduction to zero of both climate-deadly carbon emissions and eco-deadly habitat and biodiversity destruction. Honesty then demands that we ask whether our existing pseudo-democratic architecture — useless electoral systems, money buying votes, the vast power of a corporate-owned media free to lie — can possibly be adequate to the task of meeting that second demand. The fact that we even need to ask such a question is, itself, a kind of answer. Instead, we need a real democracy; we need the people to decide. And that is XR's third demand: Citizens' Assemblies to set out a path for society to follow, something politics-as-usual has so patently failed to do.

My belief — for the reasons already sketched, and on which I will elaborate below — is that, whatever you think of XR as an organisation, Extinction Rebellion's aims set out roughly the right direction of travel. We need, therefore, a broad-based mobilisation calling for such assemblies; and, failing that, creating them ourselves. This is a call that should redound way behind enclaves of left or right; the need transcends such old divides, because the emergency does. Just as, in the emergency that was World War II, food rationing was brought in by the British National Government, not as a socialist measure but as a necessity. Assemblies roughly along the lines called for by XR should be put in place at every level: local and national, but also regional, supra-national (e.g. at the EU level), and yes, and perhaps most inspiringly of all: global. They should be empowered to work in all the areas that are too difficult for 'representative' democracy to make serious decisions about. They should formulate, at last, the kinds of serious emergency responses we have been so delinquent in bringing forth.

This third and final aim of XR's could actually be the best friend of any politicians savvy enough to see why. The French President Emmanuel Macron, facing the semi-insurrection of the 'yellow vests' in 2018, seems to have been smart enough to get this. He created a national Citizens' Assembly on climate to decide on matters that, in the existing political situation and system, had become politically toxic.[87] This 'Climate Assembly' has come up with remarkably bold plans that will help Macron exceed what electoral democracy has so far enabled.[88] It has made 149 recommendations, and Macron has pledged to enact 146 of them, including seeking to bring in a Law of Ecocide, that would make the killing of ecosystems a criminal offence. This process has by no means been exactly what Extinction Rebellion wanted — it still doesn't outsource enough power to citizens to decide, and so it doesn't go nearly far enough — but it is an encouraging step in the right direction.

Politicians are deeply reluctant to really tell the truth and lead with bold action on climate; they fear that the people aren't with them and will punish them at the polls. Citizens' Assemblies could enable an end-run around that fear. When a group of citizens, much like a jury, is confronted with the full evidence, and given a chance to deliberate and reflect at serious length, they can make a decision which might be unpalatable or surprising but will almost always be respected, and will resonate with wider society as legitimate. If a Citizens' Assembly comes out with truly bold, adequate proposals to address the climate and environment emergency, the government that empowered it to make those decisions can plead plausible deniability: it can turn around to its citizens at large and say, 'It's not *us*, the government, who are saying that (for example) we need a huge real Green New Deal[89] and to cancel all airport expansion; it's you, the informed citizens.' Citizens' Assemblies can thus make bolder and better-informed decisions than referenda or representative elective democracy are ever likely to make. Citizens' Assemblies can be the people going boldly where no 'leader' has gone before. And because it is citizens themselves — rather than experts or elected (or for that matter unelected) leaders — who are leading the way, they are more likely to attain that vital democratic buy-in. Consent. Any remotely thoughtful or open-minded citizen can reflect, 'If I had sat on that "jury", and heard all the evidence and discussion that they did, I would probably have come to the same decision.'

Citizens' Assemblies, of course, will be a human endeavour, and like any human enterprise they will have foibles and failings. For a fine fictional picture of this complex reality, there is nothing better than Ursula le Guin's novel *The Dispossessed* (1974). In this book, le Guin portrays a dystopian planet and its utopian moon; the rulers of the planet, when faced with a recalcitrant anarchist revolution, eventually allowed the revolutionaries to settle their moon, and thus to avoid endless civil strife. But what is so clever about le Guin's depiction of the anarchist utopia on that moon, Annares, is that — unlike virtually all other utopias — it is not depicted as a perfect or as an achieved state. What turns out to be utopian about it is that it always needs to be *in the making*. The citizenry of Annares has to keep working at creating a viable anarcho-syndicalist society, with good-faith citizens' meetings to address new problems as they arise, including problems of ossification of the system into bureaucracy and boredom. Arriving at Citizens' Assemblies wouldn't be the end of our troubles. But they offer a mechanism for addressing those troubles that our 'representative' democracies clearly just aren't capable of resolving. A way of creating common understandings, just in those cases where doing so is hard.

Citizens' Assemblies, while coming with no guarantee of salvation, are our best shot at rising to the challenge we face. They have been XR's third and culminating demand for good reason: there will be no sufficient answer to the vastness of the ecological crisis without broad citizen consent. There isn't real democracy at present. Democracy, as Gandhi might have said, isn't an already-achieved project: it's a good idea which we ought to move closer toward. Citizens' Assemblies are just such a good idea. They will give us a truer democracy — and are likely to give us much better-informed and bolder decisions. Ireland, for example, has already had great success in using Citizens' Assemblies to make decisions about politically toxic issues; constitutional reform, gay marriage and the liberalisation of abortion laws were all achieved through this mechanism. Our best hope of pulling together and meeting the demands of this emergency is to be found in this humble and innovative idea. The time of the Citizens' Assembly has come.

What exactly will these Citizens' Assemblies decide? We don't know. That's the whole point. They will be an exercise in the wisdom of (thoughtful, deliberating, expert-advised) crowds. We can hazard some guesses. It would be very surprising if they did not plump for a massive investment in genuinely green energy.[90] And in forms of green, local, public transport such as cycling. But when it comes to other, more challenging matters, including some that have been taboo or toxic previously, we'll have to wait and see.

In the common crisis that most marks British memory, a central successful policy for our survival and flourishing was food rationing, so would a Citizens' Assembly institute carbon rationing (or indeed, bring back some form of food rationing as a precaution to hedge against eco-induced food shortages and to promote health and greater equality into the bargain[91])? Would they recommend a heavy emergency wealth tax to fund the green investment measures needed? Would they discuss the seemingly 'taboo' question of population levels — in a sensitive way, and would the outcome of their discussions issue in policies helping to consensually reduce those levels? For it is pretty obvious that the unprecedented numbers of humans on the planet come at the expense of unprecedented reductions in the numbers of other beings.[92] (Though Citizens' Assemblies looking at this will also hear about the also-pretty-obvious fact that it is mainly the footprints of the rich that stamp on the faces of the other animals of our world. Thus, it is the populations of rich countries that should be of most concern to us; and especially the sub-population that is the rich themselves.)

What you *can* be pretty sure of is that Citizens' Assemblies are much more likely to ask the hard questions and to take such radical and necessary steps than our existing institutions, and that, if they did, those policies would gain much wider assent than is currently possible. Citizens' Assemblies offer a new model for democracy itself, one suited for this otherwise cataclysmic era of climate breakdown and extinctions. Their great task will be to lead the way in our society's revolutionary adaptation to the reality of the world we inhabit, transforming us and our societies in the process, and mitigating as much as we can, along the way, the dire damage we've already done.

This is how the reworked democratic ideal focal to Extinction Rebellion, it turns out, is suitable and intended for pretty much everyone. If you have been taught to think of movements such as XR as inevitably speaking only for or to a particular 'militant' segment of society, think again. XR has sought to be a broad-based movement with demands that transcend ideology and party politics (I sat on an XR panel alongside Stanley Johnson, the Prime Minister's father, both of us wearing XR badges). XR and Fridays For Future just want ordinary people like you, citizens of what is supposed to be a democracy, to have a chance to respond adequately to the long emergency we've all been throw into. These movements aren't intended for a minority. XR — and broadly similar movements in other countries, such as the Sunrise movement in the USA, and the heroic movements that in the Global South have been resisting ecocide directly for decades — could be for anyone and everyone who loves their kids, or just cares about the future. (The challenge that I will turn to in the final portion of this book is starting out on making that dream into a reality. If you are someone who has not so far got involved in such movements, then I am looking to see how a movement can be made expansive enough to include you.)

Preparing for potential collapse too?

But we are still very far from achieving anything remotely like XR's aims. And will Citizens' Assemblies, even if we manage to achieve them, be enough? Or would elite and/or 'populist' resistance stop them from effecting a deep enough transformation? This book has been uncompromising about the gravity of the task facing us, yet hopeful that we might yet be bold enough to carry it out. But being uncompromising means being willing to acknowledge that such boldness may not prevail. So, I need to take a moment now to look even deeper into the dark heart of our current condition. It is too late in the day, the situation is too dire, for us to shy away from the possibility that we may well fail; that the slim chance we have of saving something like our world may well not get taken, despite my and your best efforts. It would be entirely irresponsible of us — it would be reckless — not to prepare for such potential failure. Just as it is reckless not to be willing to contemplate what you would do if your house burns down. If society disintegrates *without* having prepared for the possibility, the collapse will be far worse and more terminal than if we had at least tried to prepare for it.

You may well not want to hear this. You may not want to contemplate this bit of the picture at all. But just insofar as you don't want to, you need to. We don't just need Citizens' Assemblies to be empowered to define the solutions that our government and broader society need to undertake to prevent or mitigate ecological breakdown. I firmly believe that we need them to be empowered to cope with — to try, as much as is possible, transformatively to adapt to[93] — the crises that are already here and those that might well be coming.

This line of thinking suggests that the local Citizens' Assemblies will be particularly important. For they prepare us for the possible eventuality of societal breakdown. If that happens, then long supply chains and other global interdependencies will obviously be drastically reduced, which, in very practical terms, reduces the extent to which vital global cooperation will be possible. Local Citizens' Assemblies are a preparation for that potentiality. And for holding onto what we can if and when things fall apart.

If you still find yourself resistant to even considering this outcome seriously, then recall what I briefly mentioned in Chapter 1: we are involuntarily committed to further worsening our shared plight. If, impossibly (and unwisely), we were to literally stop all carbon emissions tomorrow, things would carry on spiralling downwards climatically and ecologically for a considerable time to come (especially in our oceans, because acidification will go on even after carbon emissions stop). This is, you will recall, because of the time lags built deep into the climate system, and the vicious feedbacks we know we have already set off — not to mention those we have yet to learn about. Given that we are not going to stop all climate-deadly carbon emissions and all habitat destruction tomorrow (and that in some parts of the world leaders are heading resolutely in the wrong direction), things are much worse still.

If your reaction to this dire news goes the other way — if you think, 'Ah well, maybe it's just over for the human adventure then; maybe we should just work on accepting that this is the end' — then I suspect you haven't really thought it through yet. You haven't thought, for instance, about the mind-stoppingly vast suffering that would be involved in collapse. You have forgotten what we learnt in Chapter 2, about all that our kids are to us. About who we really are.

You need to acknowledge this possibility — some would say, this probability[94] — of collapse as very real. Only if you acknowledge it in its full reality will you take it seriously enough to try to stop it from becoming certainty.[95]

If you haven't taken time to grieve this possibility, to face despair, then now might be a good moment. It is too awful to contemplate, what I am saying here, yet contemplate it we must. The only way we get to avoid this fate of civilisational collapse is if we face reality, and then transform everything, fast: the first task of Citizens' Assemblies is to map out that transformation, as sketched above. But can you honestly say you know that the necessary transformations will happen? Of course, you can't. Take some time to absorb the horror: it would be irresponsible not to prepare for potential collapse. We cannot possibly be sure that it isn't what's heading our way.

We can't keep sticking our heads in the sand because we've left things so late in the day. We have no alternative now but to try the uncomfortable feat of riding two horses at once. We need to mobilise determinedly in a last-ditch effort to prevent civilisational collapse[96]; and we need to prepare against the risk of failure. If we don't make deep and wide enough changes to stop what is coming, if the white swan of climate breakdown takes down our future, then we need to have put some insurance policies in place. And financial pay-outs will be meaningless if society collapses (the financial system being likely to be one of the first things to go down). No, the 'insurance' we need consists of measures to adapt us as deeply as possible to the ferocious changes which may be heading our way. Once again, the endeavour to make ourselves less fragile, more resilient and safer, points firmly in the opposite direction to gambling recklessly. To refuse to countenance that we may fail to stop a collapse from occurring is now, tragically, a reckless gamble. The caring, prudent thing to do is to invest some of our effort in what is called 'deep adaptation': adaptation to a possible future in which the basic structures of national and global society as we know have evaporated, as so many civilisations have done before us. Examples include reducing our reliance on sea-level infrastructure (because in any case it won't be defensible against sea-level rise, if society is failing), and re-localising our supply of food and other essentials (it is absurd, as climate chaos grows, that the UK grows only half its own food; we have caught a glimpse of that absurdity, in the sudden vulnerability we felt regarding food and other essentials, in the run-up to the coronavirus lockdown). A further example that I've already considered (in Chapter 2) is worth recalling here for a moment: rendering nuclear waste and nuclear power plants safe (because the last thing we should bequeath to future generations if they are scrambling for survival in a chaotically warming world is nuclear power plants melting down, spent nuclear fuel rods boiling their cooling ponds dry and catching fire, etc.).

Just as Citizens' Assemblies are needed to decide how to try to mitigate the climate and ecological emergency, so they are needed to help us face the possibility of collapse and start preparing for it. That's the task of deep adaptation.[97]

And perhaps the current coronavirus crisis is giving us a timely clue to this, helping to make the task real to us. It has given us a collective experience of shared vulnerability, a mass brush with mortality, a felt emergency (whereas for too many of us thus far the vast endless climate and ecological emergency hasn't *felt* like one). And, like the looming threat of climatological catastrophe, coronavirus is not going away any time soon either: it's an illusion to think of it as a passing thing. Even if there is a safe vaccine gradually being deployed at scale by the time you are reading this, there is no guarantee that it will deliver a lasting immunity to a virus which can mutate at any moment (and has already done so more than once[98]). And in any case, if we don't get eco-emergency under control, there'll be another pandemic along in a minute... Adaptation is a task that will last.[99] A lasting task for a culture re-designed *to last*.

The emergence of long-term thinking for our polity: the case for Guardians for Future Generations

Extinction Rebellion has at the heart of its vision the inspiring concept of a 'regenerative culture'. A culture that heals rather than wounds. A culture — which XR has tried (and failed and tried again better) to embody within itself — that reduces the likelihood of burnout and seeks to build and model a community that is resilient. A culture ready to survive whatever is coming; and, moreover, to have a good time in the process... Such a culture, one that would regenerate itself and help to regenerate the Earth, needs to have at its core a vision for how to build long-term survival and flourishing into its DNA.

Once the Citizens' Assemblies are well and truly underway with their work of instituting an effective emergency response, we must urgently segue to just such a true long-termism. We need to find a way of keeping our eyes on the distant temporal horizon, a perspective that has been catastrophically neglected over the past few centuries of dangerous and reckless 'accelerationism'.

I have a suggestion for how we can do this: a proposal to end once and for all the chronic culture of short-termism that blights our politics, our media, our business and economies. And when one is trying to think on a timescale of hundreds of years, or thousands of years, or even hundreds of thousands of years — which is the timescale needed for thinking (and dealing with) nuclear waste, possibly our longest-term toxic legacy — then the kind of short-term cycles that preoccupy our society of the short-attention-span don't make a lot of sense...

The very concept of democracy itself is my starting point. What does 'democracy' mean? Etymologically, 'democracy' means 'the people rule' or 'the people govern'. Reader, do the people govern in our society? Once again, to ask the question is of course, sadly, to answer it. But imagine that we got a Citizens' Assembly. Imagine that it agreed on the emergency response necessary for our society to survive, and even arrest, climate breakdown and the eco-emergency. Wouldn't that mean that democracy had arrived, that the people now governed?

But, even after this course-correction, we may still have a media that mostly doesn't tell the whole truth, a system too dominated by the interests of money, a House of Commons with the world's most pathetically unrepresentative electoral system, and a House of *Lords*... enough said; we'd still be far from real democracy. Imagine bigger still, then. Imagine that we get major reform of media ownership, the electoral system, campaign finance, and the Upper House; add in a revival (including a refinancing) of local government and a democratisation of the workplace... Are we there yet? Is that democracy, the people governing?

Even if all those changes occurred, we would still likely be living, on the whole, in a short-termist society. Why? Well, the democratic institutions that we have — even the laws that would be brought in if we made all the democratic changes that I've just mentioned — tend to be focused upon the interests and wishes of *present* people, people who are alive today. They and they alone are the people who vote and whose votes count, even in an improved and enhanced democracy. And recall the argument of Chapter 2: we need instead to take seriously the needs, the volitions of future people; what *they* will want. A people is not something that exists at one instant only; a people is something that exists over time. It begins in the past and goes on indefinitely far into the future.

On this specific point, I'm somewhat in sympathy with Edmund Burke, a 'thought-leader' of Whiggery and Toryism, who wonderfully declared that society is 'a partnership not only between those who are living, but between those who are living, those who are dead, and those who are to be born'.[100] Some of Burke's thinking is repugnant, especially his reactionary apologism for rigid hierarchy and stark inequality (most famously expressed in his defence of the *ancien régime* against the Revolution in France). However, his refusal to restrict society to the present moment is visionary, and much superior to the fixation on the interests of the current generation that is to be found in the work of his *bête noir*, the (otherwise) equally great Tom Paine.[101] But in practice, as I've hinted, Burke leans toward the past, Paine toward the present. I, on the other hand, emphasise the need to represent adequately the future. A real representative democracy would include voices that aim to represent the past and, most important, *the future*.

And while people in the past are relatively hard to harm, because they've had their time and it cannot be taken away from them, people in the future are extremely easy to harm, and indeed (in the extreme) to prevent from existing at all. Whereas, if we get things *right*, we will be giving as-yet-unborn future generations the chance to flourish, and to go on flourishing into the distant future. So, I argue, we need to find a way of making democracy actually include future people. For real long-termism to flourish, we need to find a way of representing them in our political system.

So, what would this mean? Can you give future people a vote? Well, obviously, that's not very feasible... No more feasible than it would be to give whales or wolves or cats or dogs a vote. But just because you can't give those animals a vote doesn't mean you can't try to imagine yourself into their shoes. Non-human animals have a perspective, they have preferences. Things matter to them. And we can come to understand more about this perspective: for instance, using (or, easier, just reading about) the astonishing methodology of Charles Foster in his book *Being a Beast* (2016),[102] in which for a while he sets out to *live as* a badger, an otter, a deer, an urban fox... Or ask a dog 'owner'. In some ways, namely temporally, future people are more distant from us than our companion animals. (See on this the 'How to imagine yourself into the future' exercise, below.) But in another way, it should be much easier to imagine ourselves into future people's shoes. As I argued in Chapter 2, you can make a great start by simply imagining how *much* they are likely to have in common with us in terms of their fundamental needs and values. So, let's do it. We need to find some form of surrogate representation for them. They need to have something like a proxy vote.

And think about this: so long as we don't screw up so badly that we stop them from existing altogether, over time there will be far more future people than there are present people, which would mean in a *democracy* that they would out-vote us every time. They would be the vast majority. Think about the exciting, giddy, difficult implications of this: if future generations were adequately represented in our democracy under a broadly majoritarian framework, people living today would become a tiny minority and our interests would be just one small component of a far larger landscape. But, of course, we have to be careful about that as well; a true democracy is not simply about majorities, but about people together somehow arriving at (or at least seeking out) decisions that are best for everyone. I believe that if we could imagine future people sitting beside us, thinking along with us, beseeching us to include them at every turn, we would come up with ideas and plans for how to live that would reflect the logic of this book so far. We would foster a truly long-termist culture, a civilisation committed to preserving life (and, for the reasons set out in Chapter 3, not only human life). It is, however, entirely possible — indeed, overwhelmingly likely — that certain groups will lose a lot of their privileges, for the greater good, in the course of this process. I am speaking here of those who enjoy such mind-blowing fortunes that they entertain fantasies of living on Mars or cryogenically preserving themselves for revival in some super high-tech future. Does anyone seriously think that these gross inequalities or technological-salvationist delusions make for the likeliest long-term good for future generations at large?

Future people cannot sit beside us and decide along with us. So, in order to express their proxy 'vote', I suggest that what we actually need to give them is a proxy *veto*. Because, if they were able to vote *en masse*, they would, as I say, massively *out*-vote us — unless, of course, we so badly damage our climate and ecosystems that these won't be able to support much human life in future.

Barring this extreme eventuality, however — the very eventuality that above all we have to prevent — future generations are likely vastly to outnumber us, meaning that their vote can be expressed as a veto. So I want to suggest that we need proxy representatives for future people, proxies who would be empowered to make representations to us about the interests of future generations, and to veto things that *we* might want to do but that are not in the interests of our descendants. With a nod to the philosopher Plato, who believed that we should be ruled by guardians, 'philosopher-kings', I call these representatives *Guardians for Future Generations*.[103]

A couple of questions now arise. Who should these guardians be? How should they be selected? Well, it doesn't make any sense for us to vote for them, because they are proxies for future people — they're there to express the votes that future people would most likely cast if they could. I suggest that, actually, all of us and none of us are equally well positioned to be proxy representatives for future people. As such, we need to draw these proxy representatives from across the *entire* population. I put it to you that the only fair, reasonable and democratic way of doing this is through the same principle that animates the jury system and (as we already saw above) Citizens' Assemblies — which is random selection. This system would guarantee that anyone and everyone would have an equal chance to be one of the guardians for future people. So, we need a super-jury, drawn from the entire population at random, to represent to us the interests of future people, and to exercise a proxy veto preventing us from making decisions that might harm them. The super-jury of Guardians for Future Generations would sit above our existing political institutions and have the power to veto proposed legislation, or force a review of existing legislation, and perhaps also to initiate new legislation. Crucially, they would have the power to reconsider any policy that — based on their deep deliberations, based on their task of upholding the basic interests and needs of future people, based on the absolute best expert advice available — they judged to adversely affect the fundamental interests and needs of future people.

I'm describing, then, a special Citizens' Assembly — one that would *represent* the voices of the future. Such assemblies, such gatherings of 'guardians', would be constituted at local, regional, national, international and global levels. This once again raises the question of how all these assemblies would interact with each other. The principle that should animate their interaction — especially given the risk of the collapse of the globalised civilisation we have made — is, I believe, that of 'subsidiarity'. This term means that the default option should be working at the most local level feasible. We should use the coming of these assemblies to return power to more local levels, to escape the overly centralised (and excessively globalised, hypermobile) world whose weaknesses have recently been thrown starkly into view by Covid-19.

The seminal Brundtland Commission Report was supposed to usher in so-called 'sustainable development' back in 1987 stated the structural issue thus: 'We borrow environmental capital from future generations with no intention or prospect of repaying. They may damn us for our spendthrift ways, but they can never collect on our debt to them... [T]hey do not vote; they have no political or financial power; they cannot challenge our decisions.'[104]

This is the fundamental undemocratic lack of justice I have been remedying in this section. The fantasy of 'sustainable development' did nothing for future people; but Guardians could. At every level, from the ground up, Guardians For Future Generations will seek to evoke — will in a way *be* — the voices we need to listen to, if we are to realise the kind of transformation in our attitude toward the future that this book demands of us. Going beyond an emergency response, an institution such as the super-jury of Guardians for Future Generations embodies exactly the kind of vision (and power) that we will create... if we are serious about our care for the future.

How to imagine yourself into the future

But, you may ask, how on Earth can that be done? How can we possibly place ourselves imaginatively in the position of unborn future generations, in order to represent *them*, rather than merely pretending and still tacitly favouring *ourselves*? Part of the answer was already present in Chapter 2. We — anyone — can do it, by recognising just how fundamental our commitment to the deep future is. By virtue of being humans, primates, mammals, we are capable of recognising the actual meaning, over time, of deep care for our own children. We must also recognise how reckless it would be if we neglected to apply that care not only to humans across time and space, but also to what humans need to be safe and flourishing, which (as we saw in Chapter 3) is rich, resilient, restored ecosystems. Life.

I hope you find that vision inspiring. But you may also find it abstract. And certainly, what I was asking of the reader earlier in the book was challenging.

So here comes this chapter's imaginative exercise. A way in to thinking like a future person.

This exercise is based on the Widening Circles practice shown to me by my teacher Joanna Macy.[105] It is a fairly simple practice but challenging — and the effects of it can be profound. For it to be most effective, do it not by yourself but with a small group of one or two people who you trust to listen to you (and you to them). But either way, do take a little time to do it:

Pick an issue with major long-term ramifications: perhaps fracking, rewilding, nuclear power, or the attempted creation of artificial life. Take a couple of minutes to give *your* views or intuitions on the issue. Then close your eyes and contemplate quietly for a minute or so.

When you come back from that meditation, switch: spend a couple of minutes sincerely doing the absolute best job you can of presenting what you imagine to be a *contrary* view to yours on the issue. Make sure you do so in the first person. (If you take this part of the exercise seriously, you may well find a remarkable opening or freeing up of your mind. This is valuable in itself. It also sets you up for the next — even more demanding — parts of the exercise.)

Close your eyes and meditate calmly for a minute.

Then switch. Try to take up the point of view of an other-than-human being whose interests are affected in some way by the issue in question. There are many possibilities here (and many complexities): if the issue you chose was rewilding, imagine yourself into the perspective of a lynx perhaps; or if the issue is nuclear, try to envisage the point of view of (say) a genetically damaged wolf living free of most human interference in the ruins of Chernobyl. (The real point is not what you come up with from that perspective; it is to make the effort to attain such a drastically different perspective, and the emotional change that may accompany it.) Speak from the perspective of the being you have chosen; don't just speak about it.

Close your eyes and meditate briefly again. Allow any emotions that you feel to be present.

Switch for the last time: seek to take up the point of view of a distant descendant of yours, someone born after you have died, someone you will never know: Let's say, your great-great-grandchild. Take a couple of minutes, or longer if you need it, to tell your listener(s) what you — i.e. your great-great-grandchild — feel and think about the issue in question. Speak as hesitantly or as forcefully as you need to. Look into the eyes of your listener(s). Try to connect with them, to reach them. You are speaking back from the future to now. What do the people alive now need to know about what '*you*' need? How can you best and most honestly reach them?

Then, take a final pause, but this time, rather than closing your eyes, keep making eye-contact with the listener(s). And see what happens.

This Widening Circles exercise is not, of course, intended to come up with policy recommendations, or anything as concrete as that. It rather enables one to start to inhabit the other-than-human world, and to inhabit the human future (the world after you). It starts to place one in a better position to really care for and even to *represent* that world and that future. In that way, it's surely an ideal empathetic preparation for the demanding task of guardianship for future generations. What better means could there be of preparing the way for (say) the restoration of a healthier natural world, than by imaginatively contemplating (not just from your point of view, but from the point of view of the future itself) what future beings would ask for.

If we start to hear the voice of the future, everything changes.

Acting in an emergency, thinking toward the horizon: the case for a precautionary approach

We've seen how Citizens' Assemblies could enact an emergency response to the calamitous crisis that our species (and especially our elites) have created. And how a special kind of Citizens' Assembly of Guardians for Future Generations could reorient our democratic processes towards the long-term protection of the Earth's capacities to support life and prevent the reversal of any progress to forestall the climate crisis.

Both of these much-needed proposals are a way of caring for ourselves and our descendants, of being care-full — which is both the expression of our love and true common sense. Such carefulness consists in taking the necessary precautions to stop destroying ourselves and our kin, starting with our kids. Precaution is the opposite of recklessness. The politics and culture that we have right now embody (in their leading institutions and ideology) a shameful recklessness.

There is a philosophical framework for thinking through exactly what is wrong with our approach to the climate and how it might be put right: the Precautionary Principle.[106] It is a precept that has threaded slightly surreptitiously through this book, including throughout this chapter. Consider, for instance, my insistence that we need to start immediately the massive, psychologically and logistically challenging task of deep adaptation. I am advocating preparing for an uncertain prospect which we may (we hope) never have to endure: complete civilisational collapse. The motivation for this is simple: that *unprepared is unprecautious*.

The Precautionary Principle underpins the second demand of Extinction Rebellion too: XR called for biodiversity-destruction and climate-deadly carbon emissions to be stopped by 2025 in rich countries, by 2030 worldwide. Only an eye-wateringly demanding date along those lines — impossible (within the bounds of politics-as-usual), and realistic (about what we actually need) — gives us some margin for error, some room for manoeuvre. If the massive uncertainties about the progress of the climate crisis don't turn out in our favour. And it increasingly seems that they won't; the rate of ice melt at the poles,[107] for example, has already exceeded what were supposedly our *worst-case* projections. Maybe we can survive without coral reefs; maybe we can manage without many insect species; maybe we can somehow prop up our civilisation through ice-free summers. Is it really smart to take the gamble of finding out? Isn't it wiser by far to err on the side of caution?

Moreover, Citizens' Assemblies need some way to ensure that they don't propose 'solutions' that end up making things worse, such as 'geoengineering', the hubristic attempt to control the entire planetary climate. The Precautionary Principle, I argue, is an effective decision-making tool for preventing this. And it is a Principle that would almost certainly be called upon by the 'super-jury' I've just described, the Guardians for Future Generations. We have already encountered it in Chapters 2 & 3. It is time finally to take a few moments to address it and expound it directly.

The Precautionary Principle is already observed in international law and in some national law, but it needs strengthening. (Right now, at time of going to press, powerful interests in the UK and USA are conspiring to weaken or eliminate it in the trade negotiations going on between the two countries.) If it were made the constitutional basis for our societies — if a precautionary approach were to inform everything that we do, forming the backbone of our decision-making (as tacitly it has done in this book) — it would *safeguard* us and ours.

As I understand and teach it,[108] the Precautionary Principle states: where a path of action or inaction involves a serious, irreversible risk of damage, then that path should not be taken if there is any alternative available that does not incur such a risk. Crucially, this precept should be observed even if the evidence that that risk will occur is *not conclusive*. The reason for this is that by the time all the evidence is in, it could be too late. It is not acceptable for a company to say, for example, 'The evidence that our product [cigarettes / fossil fuels / pesticides that may badly harm bee colonies] is dangerous is not decisive; more research is needed.' Rather, where the stakes are high, the onus should be on them to show that their product is safe.

The Precautionary Principle takes seriously that *nothing* is worth risking everything for. If an action could result in catastrophe, it just isn't worth the risk, no matter how slim the chance. When the outcome of taking a certain path could be ruinous, the burden of proof for showing that it won't be should lie with the corporations, scientists, economists or politicians who want to take that path, *not* on the rest of us — let alone future generations. They will probably be the ones to pay the most awful price if the gamble doesn't pay off. We can't allow recklessness with our kids' lives to go unchallenged; and they of course have zero power right now to stop present recklessness. Wherever there is threat of massive (or total) destruction, it would be careless and even unforgivable of us not to forcefully oppose it. We have to exercise the great responsibility that comes with great power, power over the future.

Human-triggered climate breakdown is an example of a ruinous threat, i.e. a threat that could cause serious and irreversible damage on a wide scale. Mass habitat destruction leading to mass extinction is another. With regard to such calamitous threats, the Precautionary Principle is decisive; it outweighs all other considerations including short-term or selfish interests.

The call for precaution again transcends the way we have been end-lessly, tediously taught to think about politics (as a struggle of 'left' vs. 'right'). It pits those of us who are willing to look critically at the way we live now, for the sake of a future, against those of us who fantasise about 'fully automated luxury communism'[109] and against those who fantasise escape from this Earth / from this mortal coil for themselves and their fellow super-rich.[110]

Creating the institution of Guardians for Future Generations would be an application of the Precautionary Principle: for, obviously, it would require of us more long-termism. Moreover, the Guardians would likely strongly adopt the Precautionary Principle. The Guardians are a democratically motivated mechanism making it more likely that the human future will exist. I don't know what they would conclude; again, that's the whole point. But I suspect that the weighty responsi-bility given them, and the requirement to look truly long-term, would make it highly unlikely that they would do anything reckless. The Pre-cautionary Principle exists to counter our tendency, especially within the short-termist profit-hungry system we've inherited, to be reckless.

Picture how different the world would be if the Precautionary Princi-ple were embedded in our constitution, if it were the first test we applied to any course of action with possible long-term consequences. Take a minute to imagine what things would be like if institutions throughout society acted precautiously, rather than recklessness being *de rigeur*.

We should recognise — as did Socrates, the founder of Western philosophy as we know it — that the wisest path is often to admit that we are more ignorant than we like to believe. We should accept that we live in a world that, in many ways, we do not fully understand. We should learn to live in a world of varying levels of ignorance and uncertainty, rather than harbouring hubristic, impossible ambitions for 'total' explanation and mastery. We should acknowledge when there are or might be insufficiencies with our models and evidence.[111] Where the stakes are high, we should err decisively on the side of caution. It is precisely for living in such a world that the Precautionary Principle is designed.

Consider a couple of further examples:

We have had an object lesson recently in what happens if you are precautious — and what happens if you are not. When Covid-19 erupted onto the scene in early 2020, myself and my colleagues, such as Nassim Taleb (of *The Black Swan* fame), argued for a swift precautionary response to this unprecedented situation[112]; unprecedented, because we had never had such a contagious and virulent global pandemic before in the age of globalisation. It was obvious that there was a very real risk of massive disruption and loss of life, because coronavirus was travelling at the speed of a jet plane (rather than at the speed of a steam ship, as Spanish Flu had done a century before, in the last roughly comparable event). It was vital to move against it *before* it arrived; *before* we knew how deadly it would be. By contrast, a wait-and-see approach (which was essentially what the UK government adopted) was reckless. The coronavirus crisis was controlled in countries like New Zealand because they swiftly instituted strong measures for pre-empting the virus's spread (including closing their borders and strict quarantine measures), judging that preventing a mass outbreak outweighed the immediate economic downside of those measures. You only get one chance at preventing a pandemic. That means that it's worth throwing nearly everything at it, even if you do not yet have all the evidence (as, of course, you don't). Take masks, for instance. Some scientists and politicians argued that masks should not be recommended to the general public until they had been proven to work: a schoolboy error. It was worth seeking to prevent the spread of this virus through masks before they were proven to work. It could have given the world a head-start on the virus, as it appears to have done in Czechia. If one looks at those countries that spurned precautionary measures against the virus — most notably, the UK, USA, Brazil (and Belarus) — it is hard not to notice that they have been the very hardest hit of all countries. (And it is hardly just a coincidence that Brazil and the USA are among the very worst countries in the whole world, in respect of climate and habitat damage...) Built into many of our institutions via the government's budgetary projections and cost-benefit analyses (which are used to assess planned projects such as HS2) is something called 'discounting'. Discounting means counting each year into the future as mattering slightly less than the year before. This idea is, incredibly, enshrined in mainstream economics, and factored into all mainstream policymaking. But it flies in the face of what is, for parents, elementary logic: that one would sacrifice one's (own) present, if necessary, for one's children (and their future). Crucially, this objection persists despite — or, more accurately, *because* of — the uncertainty over the future, for reasons of precaution.

The Precautionary Principle applies precisely where we are unsure what the future will hold. You might think that it makes sense to discount the future because we can't see it and can't know it. But the very fact that future generations are exposed to threats which are beyond our ken *is itself* a powerful precautionary reason for acting to safeguard them. We ought to ensure, so far as we possibly can, that they have maximally resilient ecosystems, and are not hobbled by our toxic legacies. We need to put them in the best position to deal with threats and challenges that we cannot fully anticipate. It is no good saying, 'We haven't got the evidence to know what they need.' Of course we haven't!

You can't 'fact-check' the future. We must act *ahead* of the evidence, and because of the inherent uncertainty. We must seek to exercise care that is likely to be helpful come what may.

These examples show us how the Precautionary Principle is nothing less than a game-changer, a paradigm-changer. It urges a transformation of our ways of assessing the risks that we have become systemically exposed to — most crucially, the danger of ecological breakdown hanging over us — and understanding what those risks demand of us.

The Precautionary Principle originated in Germany. The original term was *Vorsorgeprinzip*. A literal translation would be 'fore-care principle'. The Precautionary Principle enjoins action before potential hazards come into being. It projects care into the future. It's as demanding, and as simple, as that. And it's desperately needed.

For now, we can see how the considerations first marshalled in my master-metaphor — of our children as being what *we* become, and the direct route to the deep future — join fully with the considerations I have outlined in this chapter. Seeking to inhabit the needs of our descendants and taking up a precautionary approach to care for them, is a natural outcome of taking seriously our care for our own offspring.

The urgent case for slowing down

Citizens' Assemblies enjoined to face our clear and present emergency and chart a path out of it; Guardians for Future Generations empowered to protect our descendants and ensure that we never fuck up so badly again; and, across it all, a precautionary approach embedded in law, politics and everyday life: these comprise my proposed political-philosophical recipe to address the greatest crisis we've ever faced.

And so I've completed my pitch. This, I argue, is what a sane society would do, faced with the long emergency we find ourselves in. A sane society would respond swiftly and with absolute resolution. With determination, and with deliberation. Furthermore, my proposals map out a smooth transition from the necessary immediacy of emergency action to an equally necessary focus on and care for the long-term future. A long-termism that is essential if we are to escape the narrow temporal horizons that got us into this awful fix in the first place.

I started this chapter by noting that there's a seeming-paradox about our situation: that there's nothing more urgent than acting with a long view. That we desperately, urgently need to figure out how to become long-termist in our way of thinking. (And thinking is nothing without acting on the basis of it.) But I hope in this chapter to have solved that paradox: if we create a Citizens' Assembly as soon as possible, if we then institute something like Guardians for Future Generations, and if we thread a precautionary approach throughout our deliberations, then we really can start taking care of the future in its unimaginably lengthy unfolding, right now.

Plus, recall that *local* Citizens' Assemblies could be a way of rebuilding some democracy, some well-informed participatory organising-power, *from the bottom-up*. This will be a very good thing to do anyway. But it will be of profound importance in the event of some societal or state breakdown.

My final chapter turns to the most practical question of all: how can we actually achieve these political changes, enough, we may still dare to hope, to prevent such breakdown?

And at this precise moment in time and space, at this very instant in which *you* are reading these words, that boils down to: what can *you* do, reader, to help realise the vital project of deep political and institutional reform set out here? What is your calling, to do and to be, to help modify our attitude toward time itself? And to do so in time.

Time to turn to where you go from here; before humanity runs out of time...

5

What are you to do?: your money and/or your life

'This is our first task — caring for our children. It's our first job. If we don't get that right, we don't get anything right. That's how, as a society, we will be judged. [...] And by that measure, can we truly say [...] that we are meeting our obligations?'
– Barack Obama[113]

This book has been an ingenuous effort, free of the knowing cynicism that deforms our public life, to think through how we find ourselves placed, at this awesome, terrible (and, if we are willing to face it adequately, deeply hopeful) moment in our history. There is a real risk, when contemplating how close the human race has come to irretrievably fouling its own nest, of descending into what the philosopher Nietzsche called the philosopher's greatest temptation: nausea at the human race. And for sure there is a great nausea coming. As more of humanity wakes up to the extremity of our self-imposed predicament, a common reaction will be, 'Then we don't deserve to survive.'

But the truth is that that reaction and that nausea are essentially nothing but yet another excuse for inaction. If you follow instead the logic of this book, then you'll come back (instead) to love, to your children, to this beautiful living planet. These should not be tarred with the brush of our present failing civilisation. If we change things radically enough to save the future, or at least to make our descent more civil, intelligent, decent and caring, then we'll (at least) have created something beautiful and noble in the greatest of adversities. And we'll fully deserve to be treated like good ancestors — like heroes — because of it.

There can be no excuse for inaction.

This book has laid out a compelling motivation for acting and a roadmap for lasting change. Chapter 1 introduced the nature of the crisis, the long emergency, that we now face: a potentially permanent emergency that changes everything, and puts what you care about most of all, those little ones who depend upon you, at existential risk. From there, Chapter 2 showed that simply truly caring about your kids unexpectedly turns out to equate to caring for the whole human future. Chapter 3 showed that caring for the distant human future turns out to equate to caring for the whole planet and that that care begins right now. By the end of Chapter 3, then, you've seen that in order to protect our own kids from the catastrophe that is (as things stand) on the cards, we need to show a consistent care for this living Earth, now and forever. Having established this powerful motivation for acting to prevent climate nemesis, I turned in Chapter 4 to prescribing an emergency response to the present crisis and a set of measures for embedding ecological long-termism in our polity. I set out the kind of policy framework and institutions needed to achieve such caring. A framework, as one might put it, of love incarnate.

You love your children; so, it turns out you love distant future generations; you love the whole Earth so you will seek to make this love manifest in our key societal institutions. To embody it.

The task in this chapter then is to consider how you, as an individual, might best seek to embody these goals.

Consider credible best and worst case scenarios, in relation to those goals:

We get this right, and we'll find the road to a better life... We'll face worsening disasters for a long time to come — but they'll bring us together.[114] We'll have to give up much of what we are used to — but we'll gain a comparative freedom from noise, pollution, the rat-race, and from rampant growing insecurity (including crucially food-insecurity). We'll have to make unprecedented efforts and sacrifices — but they'll give our lives meaning: a meaning and a nobility that is lacking so long as we are set almost willingly to bequeath to our children a world declining indefinitely into the future.

We get this wrong, and we'll continue on the road more travelled: the road towards climate-meltdown. Here's my nightmare: one of these years, unprecedented weather chaos ruins most of the world's harvests. We manage to get through that without mass death, due to using up virtually all our grain reserves. The next year, the same happens again: and suddenly we are getting unquenchable famines breaking out, in several regions of the world simultaneously. Suddenly, even rich countries are not immune. 'Multi-bread-basket failure',[115] it's called; and each year that we go on like this, it becomes more likely.

Probably what will happen will be somewhere in between these two broad scenarios.[116] But don't let that reassure you; for right now we are solidly headed for something more like the latter than the former.

The task, the great work, is to change our consciousness so that we move the dial closer to getting it right than to getting it wrong. Every single bit of difference that gets made could mean someone's child living or dying. Or it could even mean many millions of lives. For it could mean the difference between a tipping point tipping, or not.

And reader: the consciousness in question, at this moment of reading, and ever after, is your own.

For I've sought in this book to take you, heart and head, on a journey. Each of the three main chapters of this book, Chapters 2 to 4, took the following form. I started with an unobjectionable emotional proposition: at the start of Chapter 2, it was simply that you truly care for your own children. I then drew out the logical implications of that: in the case of Chapter 2, that you are committed, therefore, to caring for the long-term human future. I then sought to deepen this emotional commitment through an imaginative exercise. This pattern of emotion-logic-emotion leads from one chapter to the next. At the start of Chapter 3, the emotional proposition became that you care for the human future generations from now. Over the course of the chapter, I showed how this logically entails care for planetary ecology, and I suggested an imaginative exercise to undergird and make vivid that commitment. By the end of Chapter 4, this process resulted in a recognition of the need to make huge, swift changes to our institutional framework, to our system; to prevent societal collapse (or at least to try to find a path through it). And so, we come to this, the final chapter of this book, in which it is necessary to ask how to *achieve* that framework. How to bring about the practical incarnation of love at large, the compassionate revolution about the fixed point of our profound care.

More than once in this book, and especially in the previous chapter, I have asked Lenin's famous question, 'What is to be done?'[117] But if you have got this far, and are more or less convinced, then the following more pointed question, not framed in the passive, rears its head, 'What am *I* to do?' For it is self-evidently not enough to pronounce on what 'everyone' or 'society' ought to do and leave it at that — especially if leaving it at that means adopting a pose of spectatorial superiority, or of irritated condescension towards those (such as governments!) who have palpably failed thus far. No, the pertinent question now is not what *ought* to happen, but what *you will do*. What, now, will be your will...

The first thing to say in response to this question swiftly takes us back toward the collective. For my first advice as to what you individually can do: don't let what you do in response to what you have encountered in these pages be restricted to the purely personal. Purely individual responses are of limited impact. If you decide to stop flying, or have less children, or what-have-you, that's super, but that alone doesn't get us far. Certainly not far *enough*. For here we should recall the words often attributed to Churchill: 'It is not enough that we do our best; sometimes we must do what is required.' Churchill wouldn't have had as much patience as I've essayed here for trying our best... In the extreme crisis in which we now find ourselves, only a swift, society-wide response will be enough. So that is what is required (as in Chapter 4). But initiating that response cannot be left up to that amorphous entity, 'society'. Initiating it is a job that falls to you and me. Perhaps you don't want the responsibility. But that is no longer an option. This book has explained the urgency of the situation and shown your duty to act. This exchange from *Lord of the Rings* might be helpful, at this point:

> **Frodo:** I wish it need not have happened in my time.
> **Gandalf:** So do all who live to see such times, but that is not for them to decide. All we have to decide is what to do with the time that is given to us.[118]

We have to decide what to do; and then do it. So the first thing to do as an individual is to be more than just an individual. Throw your lot in with others. Be clear that if collapse comes the minimum unit that can possibly survive it is a community. Be clear that the minimum unit that can possibly avert such collapse is a society. Build community; and join with others to try to bring about changes such as those outlined in the previous chapter. 'What are you to do?' becomes 'What are *we* to do?' again. So, think — and act — collectively.

If there is to be future, we need to think as a 'we'. We need the opposite of the story of separation, the fantasy of rugged individualism, that has brought us to this sad pass. Think like a super-organism, a team, an ecosystem. Think like a growing, resonating wave. Think like a movement. Think: 'What are we to do?'; and become part of that needful 'we'. Move to co-create it.

Don't hold yourself apart; let yourself grieve for what has already been lost, for the security (about the alleged endlessness of 'progress', about having a pension, about your kids having a definite future) that has been taken away from you and yours; let yourself fear; let yourself rage, even. And then morph that energy back toward the love that underlies it. And so, into *will*. Into what I'd call *the will to empower*. Empower yourself and those in the moving wave with you, those willing to will compassionate revolution.

Harvest the beautiful energy of grief and fear and rage (and even of terror and panic and depression), all of which are really at root the energy of love. For grief and mourning are the honour that love offers to what is lost; while fear and anxiety are the honour love offers to what is not yet lost; and anger and rage are among the currency love pays to stop the loss with. Once you have attended to these feelings, and let them flow, then it's time to give of yourself. To commit as much to the 'we' as you can. What can you offer to help move what needs to move? What specifically will you contribute to grow the wave? How can you help build the most important movement in history, the one thing that our children and their children will judge us by (and if we fail abjectly — if we don't even truly try — then believe you me they will judge us)?

Perhaps, in what you devote your life to, in your work or in your personal mission, you are already doing all you can, to incarnate as effectively as possible the great love, to care for your kids — which means to care for the whole Earth now and forever, and to elicit an emergency response which will manifest that care. If this is so, then you can stop reading now. But it may be wisest to consider for a moment what is truly effective. As I've already noted, personal change alone will not cut it. Shorter showers are not going to stop climate breakdown.[119] We need everything, and we need it joined-up, and fast. We need *wholescale* change: the whole system needs to change, and the scale of the change needs to match the scale of the crisis.

Can we expect this kind of change through the conventional political system? Nope. President Sanders rather than President Trump would have helped; but even Sanders still believed in endless economic growth (on a finite planet, no more!). The right kind of electoral politics is clearly part of what we need. A really significant part. But it isn't enough. President Biden barely gets us to first base. We need a massive shift in the whole political *agenda*. We need a revolution in consciousness. We need a dramatic, comprehensive shift in priorities and practices, across most of the world. Ecologism[120] is one country would be better than in none, but it is self-evidently nowhere near enough. The kind of thinking found in Chapters 2 & 3 needs to be absorbed by some of the elite; there needs to be sufficient recognition that there is no taking care even of our own without taking care of the deep future, including the non-human. The kind of proposals essayed in Chapter 4 need to become politically possible, and then politically actual. That starts with our emergency response: Citizens' Assemblies to chart a way through the emergency, and precautionary thinking and action to become the norm (rather than, recklessly, the exception).

Let's be blunt: this is an utterly demanding set of asks. This is why I think it obvious that we need movements like Extinction Rebellion and the Youth Climate Strikes. Personal change, intelligent electoral-political action, good work from non-governmental organisations and charities: these matter, very much. But in our hearts and heads we know full-well that they are not *enough*. Non-Violent Direct Action, in the school yard, in the workplace, in the streets, and descending in huge numbers on the failed pillars of our failing system — government, media, *haute finance* and big business — will be necessary to radically shift what is acceptable. And such NVDA needs to happen at unprecedented scales; much bigger than it has, yet. We have to make what is 'politically impossible' seem not only possible, but necessary, obvious. (And, if the needful change still doesn't come, then we'll do as much as we can of it ourselves: NVDA should, in such circumstances, be used to transformatively adapt our systems directly. Including crucially: many of us who wish to getting back closer to the land.[121])

The Climate and Ecological Emergency bill[122] recently brought before the UK Parliament (in autumn 2020; it will be debated in spring 2021!) aims to achieve a fully empowered Citizens' Assembly that can manifest and yield democratic citizen buy-in so that together we do enough. More broadly, movements such as Extinction Rebellion aim to achieve a sufficient emergency response, and to facilitate the emergence of a *regenerative* culture, a culture that will recreate itself, replenish our society's resources, and sow the seeds for a new long-termism. Crucially, this would include embedding the Precautionary Principle deep in policy and criminalising ecocide, the murder of ecosystems. The kinds of proposals that I set out in Chapter 4 are, for the first time, potentially, conceivably on the cards; *if* what happened in 2018-2020, with citizens (including our future citizens) stepping outside the system in great numbers, now gets massively scaled up. People are talking about Citizens' Assemblies, and even bringing them into being. Every citizen, regarding what a Citizens' Assembly decides, can say: 'There but for the grace of the lottery go I, getting well-informed, having the space to listen and reflect and discuss, deciding on something that actually works for the common good.' This is why Citizens' Assemblies tend to work, and why smart politicians will see that such Assemblies are the way that vital issues currently stuck in the 'too-difficult box' can finally get tackled.

What needs to happen is finally starting to happen. We just have to be bold enough to play our cards right; and that means focussing on changing the whole game. XR and Greta, and their allies and offshoots and potential successors, have offered a grain of hope, for the first time in years, that it ain't over yet.

This is especially so given the astonishing way in which we have *seen* everything change overnight in governments' responses to the coronavirus pandemic. Unheard-of amounts of state resources were mobilised by regimes that had told us it was too expensive to do what we'd urged to save nature and climate. Never again will it be possible for them to hide behind the excuse that they lack enough cash. Especially once it becomes more fully understood that this pandemic *is part of the ecological crisis*. It was caused and vectored by the worldwide fanatical pursuit of economic growth, by habitat-destruction, by maltreatment of animals, and by the real super-spreaders: jet-planes.

When we went into lockdown, we literally contracted the economy overnight to protect our vulnerable. And in countries like the UK, the people led on this, and the government merely followed[123]: many of us cancelled events and went into isolation for the common good way before we were ordered to. This is a hopeful precedent. The kind of protection we offered the old from coronavirus must now be extended to the young from the climate crisis. We've seen some governments (Vietnam, South Korea, Taiwan, New Zealand, and, after a deadly slow start, China) follow the Precautionary Principle[124], and their populations benefit from it. We've seen other governments (most notably, the USA, UK and Brazil) not follow it,[125] and their recklessness has resulted already in hundreds of thousands dying unnecessarily — victims, effectively, of state manslaughter. That lesson can be carried forward to the longer emergency which has been the central concern of this book. We have had a shared experience of vulnerability and emergency now.[126] Across the whole globe. That is an unparalleled gift that this deadly virus has unexpectedly seeded. This gift has a deep saving power. All that is needed now is for us to *learn* from this shared experience; that from this emergency should arise a deepened sense of what matters, a stronger determination to save our vulnerable (not forgetting ourselves!).

This communal experience of vulnerability opens up a new space of social and global possibility.

We have half-missed this possibility already.[127] Already, too much rushing around has reinstated the deadly air pollution that we live with as if it is 'normal'. Already, way too much money has been spent on bailing out business as usual.

How we emerge from the coronavirus crisis in 2021-2 almost certainly dictates whether we will take or blow our last chance at averting eco-driven societal collapse, probably during the lifetimes of many of us.[128] This is our last chance because these opportunities for radical reset don't come along often. The last one, which we blew completely, was in 2008. That's over a decade ago. A decade from now, all hope of holding global overheating at 1.5 degrees, a reasonably safe level, will be gone, unless we commence wholescale change *now*.

For the climate crisis is very like the corona crisis — only it plays out over a vastly longer timescale. We needed really to have begun to transform our system *decades* ago.

So, I am quite simply stating a fact when I say that the next year or so is the last chance. The climate timeline dictates that conclusion, as UN Secretary-General Antonio Guterres has stated; and the corona crisis dramatically underscores it.[129] The money-tree that governments are accessing now consequently won't still be there in a few years' time. There will then be calls for belt-tightening. If we don't get this reset right now, then we will be fixed on a course to crash later.

Extinction Rebellion came into being to make the 'politically impossible' possible. If it or movements alongside it or succeeding it were to rapidly become much richer and larger, then who knows what might yet be achieved. The ideas outlined in Chapter 4, or proposals like them, could become a new common sense much faster than one would have thought possible, at a time when so many are realising that the existing system is broken. And now that we've seen business as usual suspended for the first time ever, the coronavirus recovery period is a perfect time to radically rethink nothing less than *everything*.

We need such movements — including for transformative adaptation of our failing societies — operating at scale, and we need this soon, if we are to stand a chance of making sufficiently radical change. We need to act now to end the biodiversity and climate emergencies before they end us.

Maybe now you are focussed (intellectually) on the right kind of action, but not quite there yet in terms of your own *commitment* to it... and I need to say to you that this is very likely to be the case even if you think it isn't... For this crisis demands of us more than we've ever given before. (Y)our kids are on the line. Their lives and the whole future is on the line. Bluntly: everything, including that which matters to us most

(our children's wellbeing), is in line to be eaten up and spat out. V
the stakes this high, you need will and commitment like never bei
We need to rise up the meet the full scale of this mother of all cris

What comes to mind for me, in this epochal existential emergency, is the old catch-phrase of the highwayman: 'Your money or your
life.' What matters to you most of all is about to be swept away. Why
wouldn't you put your life — your time, your livelihood — on the line
to stop this? And I'm not even asking for you to sacrifice your own
existence. I'm only asking you to devote your life to the cause of Life.

If you truly can't do that (because perhaps your life is fully occupied
by more immediate cares that you *cannot* escape), or if you can have
a greater effect by donating your resources than your time (because
perhaps those resources are substantial), then fine, I'll ask even less.
If you can't give your life to this, then give your money. A more modest
proposal would be hard to find, given the situation.

But let's be clear. When I say 'your money', I don't mean a
£10-a-month standing order to the Green Party or XR. I mean your
money. You need to ask yourself with an honesty that you have probably
never yet really brought to bear: do I *need* that foreign holiday? That
extension or loft-conversion? (That second home?!) Do I need it more
than my kid (which, recall means: more than the whole human future)
needs it? If you have an ISA, or a second home or a property you rent
out, or an inheritance, then what you are doing with your life? The time
is now; we/you have to try, while there is still time. What good will that
money be to you when the banks fail (which at some point they will, if
we go much deeper into ecological debt)? You can't take it with you, nor
can your kids benefit from it if we go back to a barter economy.

Coronavirus lockdowns were a tough time for some of us, especially
if we had kids, or no garden, or a stressful and hazardous frontline job.
But I think something that struck many of us was that, actually, there's
a hell of a lot in life to appreciate without having to constantly travel
and rush around and 'make money' and spend like there's no tomor-

row. To hear the birds sing (and for once without their being drowned out by traffic or planes), to meditate maybe, to read or write, to have the time to Zoom with friends we hadn't spoken to in months, to buy food for a vulnerable neighbour, clap and cheer once a week for our carers alongside everyone else on our street... It was astonishing how much of a good time one could have with so very little. A key reason, of course, is that our lives had clear *meaning*. We were staying in to protect the vulnerable (and to protect ourselves).

Now in the longer climate and ecological emergency, which will remain long after Covid-19, long after the virus fades into the furniture, we need to do the same. This whole book has been about protecting the vulnerable. In the case of coronavirus, the biggest class of vulnerable people was the old. In the case of the climate crisis, it's the young. As I've emphasised throughout, this project — sustaining and not destroying life — is all about them. Despite the absurd intention initially of governments such as those of the USA and UK to allow most of us to get infected by the virus, and our healthcare systems likely overwhelmed, in order to allegedly achieve 'herd immunity' and to protect 'the economy' (which of course was instead hit much *worse* than in countries that acted with foresight), we managed to save most of our old during the worst of the crisis. What is now needed is the mother of all acts of *intergenerational solidarity*. We mostly managed to save the old; now they/we must save the young. That is the beautiful burden on our shoulders. The task sketched out in these pages. When the school climate strikers issue their plaintive call of 'Save our world!', it's time to *hear*. Especially (but not only) if you are old and rich or not-so-old and strong. Your body, your mind, your money, your life: the time is now for these — for you — to go into service.

Consider the Extinction Rebellion of April 2019 in London (and across the UK and internationally).[130] This revolutionised public attitudes to the climate emergency in this country I'm writing from (the UK), and led to a Parliamentary declaration of Climate and Environment emergency, to a net zero carbon target in law in this country for the first time, and to a Parliamentary-mandated Citizens' Assembly on climate. And it was achieved on the basis of a tiny shoestring (compared to the budgets of NGOs such as the National Trust or the Royal Society for the Protection of Birds), about a million pounds in total. Now that's value for money. Your grand, or your two-hundred grand, or whatever it is, could play a significant role in the next huge leap forward for climate-consciousness, the next ramping-up of ambition, that is desperately needed.

The cause I'm essaying a pitch for in this book is life itself. Giving ten thousand pounds, or whatever you can, to this most vital of causes is a small ask, compared to losing your liberty[131] — or, indeed, compared to losing your self-respect if you fail to heed your conscience and step up to the plate.

So do get on with it. Once you've opened you heart, open your wallet. Wide.

Similarly, if, rather, in your case, it's your *time*, your life, that you can donate to following through on the implications of this book, then that doesn't mean joining a demonstration once a season or so. It means getting serious. The inspiring school climate strikes can only happen if children are willing to risk the ire of their teachers or parents (not to mention politicians) and go *on strike*. Friday after Friday. For we adults, the responsibility is much greater: it is obviously unacceptable to place the burden of savings our kids' future on those kids. I didn't get to be an Extinction Rebellion spokesperson on telly or represent XR at meetings with Government without thousands of rebels willing to get arrested first. Some of us are going to have to get arrested. Some of us, like the Suffragettes did, are going to have to go to prison. Some of us will be vilified for a while by shameless elements of the press that hopelessly cling onto the status quo (though, in the longer run, if we succeed in ensuring there *is* a longer run, we'll be looked back on as heroes). Some of us may even get physically injured. Some of us are certainly going to have to give up many of our evenings to meetings that we might rather not be at. We'll need rather more determination than Oscar Wilde was able to muster.[132]

But let us find it. For the logic of the previous chapters is implacable. If you care about your kids, then being willing to make some such sacrifice is what has now become your vocation. And you know what? It's no *sacrifice*, actually. There is nothing more joyous than standing where and when one can do no other, with one's fellow humans, smiling, singing perhaps, in service to the future and this vastly beautiful living rock. Just as a parent won't hesitate to put herself between her child and harm's way, so we are the spirit of the planet, rising up in collective self-defence. We are the smoke alarm, we are the emergency response, we are the arising consciousness. Being part of this in real time is no sacrifice. On the contrary, nothing is more meaningful, nothing feels better. The subtle secret at the heart of the radical activism sweeping our planet is that taking part in it, even and sometimes especially when the consequences are personally challenging, is the opposite of hard. For nothing could be easier, in the end, than letting oneself express the love that one is. Committing oneself to active involvement alongside others in a direct action movement is joyous. It's like coming home. You're alive like never before.

This book has been all about asking us — asking you — to really try. But if that feels hard, then there's a sense in which I haven't landed the point yet. When the meaning of this book really lands, then your change of course won't feel hard at all. Rather, it's effortless. It's no sacrifice at all.

Consider one last time the astonishing, moving worldwide school climate strikes: our children are waking up. As the waters rise, so do they. We adults need to awaken too and act decisively — rather than leaving it to *them* to show all the leadership. For they are starting to lead us. But it would be an awful act of bad faith, an abnegation, to leave the future to them to save. They haven't got time to take and change power as adults and so save themselves. For it has to start *now*. Think of it this way: we're reaping the consequences of *our* parents' generation's inaction. *Silent Spring* (1962)[133] came out before I was born. The first Earth Day was fifty years ago. The great Club of Rome *Limits to Growth* (1972) report[134] — which, except for its excessive *optimism* about carbon pollution, appears to have had its finger more and more on the pulse with each passing year — appeared soon after that. Since the late 80s, it's been clear that climate change is deadly, dangerous and human-caused — and yet since then we've bunged more greenhouse gases into the atmosphere *than in the entire history of the human race prior to then.*[135] We're reaping the consequences of our parents' inaction. Let's not repeat their mistake. Let's take this last chance, unexpectedly afforded by the 'global reset' of the coronavirus crisis, to be good parents to the future.

As I've set out, this is almost certainly our *last* chance. Wouldn't it be glorious, if we took it together?

Look, I'm not here to lecture you. I'm really not. This is about listening to your heart and thinking through your options. Maybe you'll give your money, gradually; and maybe you'll find a smarter way to do it than what I've suggested. Maybe you'll work up to giving your time; and maybe you'll then make it count in ways that are undreamt of in my philosophy. Maybe you'll find some judicious or pragmatic combination of both or throw into the mix something else altogether. Maybe you are very pressed, as many people are in this place and time, and the most you can do is to support *others* as they give their money and/or their life to this. If so, great.

I'm merely asking you to reflect deeply and with integrity on where you can make the biggest contribution to this most vital of projects. Perhaps in your case it's by twisting the arm of someone very well-connected who you are well-connected to; or by standing for election. Perhaps it's by giving what is honestly as much as you can; or by putting your body on the line.

Just don't shoot the messenger. If you find yourself feeling annoyed at *me* for trying to force you to confront all this, for 'asking a lot' of you, that itself might well be a sign that you are feeling guilt that you aren't doing as much as you could be. Your kids — and all of your descendants across the whole future — are depending on you. I have argued above that the only truly effective way to act is collectively — but, reader, your role is non-negotiable. The collective has to *start* somewhere. You are part of us. You're part of this world,[136] and you cannot shift your responsibility for it to anyone, or for any price.

What I've tried to do, in writing this book, is to facilitate your *feeling* this. To share with you my passion for the primacy of the next generation — and what that turns out to mean. What have I actually achieved? Well, I'm confident in one thing at least: I have made blunt asks. I have been far more direct than people usually are; I've done the oh-so embarrassing thing, of putting you on the spot in asking for your life or (more difficult still to talk about) your money. But, of course, I don't really *know* what I have achieved. Because that depends utterly upon you. Upon the genuineness and depth (or otherwise) of your response.

This book is coming towards its end. But an essay is by definition incomplete; it is as much a promise as it is a product. In this case, the promise called into being by the book is one that has had the audacity to ask you to join me in delivering on it.

Simply put: the end of this book is not its completion. *You* are needed, to complete the vision begun here...[137] This essay is incomplete, without your endeavour, your responsive action.

I have operated throughout this book on the basis of two incontestable, conservative assumptions: that we are in the throes of a severe ecological crisis which may threaten your children (and theirs) severely, and that this (severely) worries you, because you love them. These simple, obvious, undeniable truths alone are enough to generate a care for the far human future (Chapter 2) — and, in turn, for the far ecological future (Chapter 3); this care for your own kids calls you to immediate and long-lasting action to preserve and restore the ecosphere. These twin asks, of emergency care for the present and long-termist care for the future, require radical (and rapid) societal and institutional change (Chapter 4). And for that to happen, we need *you*. So, this chapter has been about how you need to act as if you mean it. You need to act as if your life depended on it — which it does. Because the crisis is here now; this generation is already suffering, and it's going to get worse. And, more important (still), because your life in its real meaning *is* the next generation. And so *on*.

From barely noticeable acorns come mighty, inspiring oaks, built to last. Truly caring for (y)our own children, in the way sketched in this little book, can and should add up to saving the entire planetary future. Deep time delivers deep purpose... All those generations! All depending on us. All hanging by the slender thread of the decisions we make now.

What do you have to lose?

Everything. And so: we can but try...

A PROPOSAL:
Parents For A Future

So you know; and so you know what to do.

Or maybe you feel that you actually don't. And if you don't know just what to do just yet; actually, that's pretty normal. That's *healthy*. Give it time. Stay in 'negative capability' awhile. The most urgent thing to do right now (and here's a final one of those paradoxes of time that we've met in these pages) is to *pause*. So that you are fully committed, and to the right thing, when you move. And so that we don't endlessly *rush*...

But if, after taking your time, you still don't see a way to do enough, still don't see *your* way, then I have a suggestion for you...

First, recall that we were all parented. We were all children, needing protection, needing adults willing to take responsibility.

This book has been about parenting the future. About finding the courage to insist that we *will* parent the future. We *will* protect our coming generations as very best we can. We will do what it takes. And we need to do it together; not just because together we are strong. Also, because it's just too damned painful to try to do it alone.

Perhaps what we need now more than anything, to make this vision, this commitment, into a reality, once we move — perhaps what might actually work, what might be inclusive and so *big* enough — is a movement that embodies precisely this idea: parenting the future. The idea that has been the core of this book could be taken literally, in our lives: by building a new movement.

Fridays For Future, the beautiful global upsurge that has been the youth climate strikers, brought together many of the world's children to call, with extraordinary pathos, for the older generation to take care of the planetary future, now. The call has reverberated around the world and has lodged in the hearts and minds of many. (When I was in Davos for Extinction Rebellion in January 2020, telling the truth unreservedly in the citadel of the elites, the thing above all that helped get me and my colleagues something of a real, surprising hearing was a refrain we kept hearing from those elites, from corporate CEOs and the like, usually along these lines: 'My teenage daughter won't let me off the hook. At breakfast almost every Friday, she demands to know why I am still involved in the slaughter of the living planet! I don't have a good answer, so that's why I want to listen, and to help shift the business model that is threatening her generation, if I can.')

But, as Greta Thunberg has stated with her trademark blunt honesty, the last two years have been mostly wasted.[138] The world may have listened, but it hasn't acted. The elites are still a million miles from the kind of emergency-response that is required. Just contrast the mostly rapid and large-scale response to corona with the still pitiful inaction on climate and nature. And just consider how dangerously un-climate-proofed most of the post-Covid 'recovery' plans are.

And in any case, as I emphasised in Chapter 5, and as Greta has pointed out with unnerving moral force, it's just *wrong*, to leave it to our kids to try desperately to convince us to save them.[139] It ought to be *us* leading; it is already a massive fail that it has come to the pretty pass that this has somehow been left to them!

Furthermore, the most vulnerable of all — really young children, not to mention unborn future generations — obviously cannot take part in eco-action such as school strikes at all. There is never going to be a 'Toddlers For A Future'; at least, there is never going to be such an organisation led by toddlers.

So, what if, instead, we turned the telescope around? What if, rather than shamefully relying on our children to lead, *parents* were to step up to the plate?

Marx famously saw the workers of the world as the class that would determine the future. I've thought for many years that the most relevant game-changing 'class' now, in the twenty-first century, in the age of enduring ecological crisis, is the parents of the world. But until very recently, this idea wouldn't have made a deep enough sense to most people. Only now, with the climate itself starting to free-fall, and in the time of post-Covid reset; only now, at this moment of last chance for humanity, and with Greta and XR having blazed a trail; only now has the time ripened.

The movement I am inviting you to imagine with me is now possible. We are finally at the point where parents are starting, painfully, to recognise that responsibility for dealing with this existential emergency *cannot be outsourced any longer*. Governments are not going to sort it. Nor are corporations. Scientists completely lack the power. And there just aren't enough passionately active citizens to turn the tide... yet.

Imagine a global movement so massive that even Fridays For Future pales in scale beside it. Imagine parents systematically, tirelessly deluging MPs' offices and mailboxes, till the system can't hold back the tide any longer. Imagine parents refusing to take no for an answer: simply insisting, against the vested interests that will oppose tooth and claw the kind of radical measures I put forward in Chapter 4, that their children must be given the right to live, and must not be reduced to begging for their lives — as the climate strikers have been.

For a parent's love is a mighty power. Woe betide any force that seeks to come between parents and their kids' futures.

Imagine parents *going on strike* en masse every (say) Monday till governments and corporations bend (or else they break). It would be like a weekly general strike,[140] a giant gathering or march to make the stakes stark and the consequences (for the powers that be) real. There is strength, of course, in great numbers; this movement would empower many who traditionally find it too scary to move outside what government or society takes as 'acceptable behaviour' to do so. And soon it would be the *norm* for parents to be thinking — acting — this way, for a future. When it reaches that stage, Parents for a Future will be 'mainstream'. (There will be no way of 'ghettoising' it the way that unpleasant Establishment or hard-Right media tend to seek — not unsuccessfully — to marginalise 'spiky' movements such as XR.)

From that solid platform, maybe some adult strikers would feel moved furthermore to engage in very civil (but very serious) disobedience, perhaps against those who are most responsible for the wilful eradication of the future: such as the Murdoch empire, Justin Trudeau's sponsorship of the tar sands in Canada, or Brazil's Bolsanaro in the Amazon and those who finance the demolition of this great lung of the world. And probably also for what's needed instead: for a just transition from the dying fossil economy to an economy that works for people and planet, maybe for a law of ecocide or Guardians of Future Generations... If even a fraction of the movement we're now imagining were willing to do this kind of thing, it would result in non-violent direct actions that would make XR's Rebellions look tiddly in comparison.

For imagine the parents of the world — imagine tens, then even *hundreds of millions* — uniting in their pain and their love and their desperate active hope. Imagine the biggest march in human history, timed possibly for the day after the near-certain failure of the global climate COP conference in Glasgow, in November 2021...

Vision in your mind's eye the unanswerable refrain as, one after another, parents say into the cameras, 'I'm doing this for my daughter, and for *her* children, and for their children.'[141]

Dare to imagine all this, and you're imagining Parents For A Future.[142] You're beginning to imagine us finally doing enough.

What exactly will this movement call for? We can't know. (Perhaps it will create Peoples' Assemblies of its own to decide.)

But here is an essayistic guess: I think the parents of the world will increasingly want the world to move toward the kind of action on the climate and on nature that Fridays For Future and XR have called for. But I think they are likely to emphasise more explicitly the need to redesign the economy so that it functions in a manner that isn't crazy. Parents who pause and then move together may insist upon worthwhile, non-harmful livelihoods for their children. There's no jobs on a dead planet, but, equally, no future without decent jobs. I think Parents For A Future will urge us all to face the question of livelihoods that make for a sane society; beginning with such essentials as growing food, doing caring work, producing and fairly distributing genuinely renewable energy.

I think they'll wants us to explore how to translate the care highlighted in Covid into care for the future.[143] This could mean societies themselves being redesigned so as to place care at the heart of everything we do, everything we are. Care in the sense of emotional care, love; a deep care for physical and mental health; and care in the sense of care-fullness, precaution as opposed to recklessness.

I think they'll press the question of how to take adequate action on these matters in a way that draws upon the needs and wishes of us all. Which means they will probably call for — and where necessary, where the call is ignored, co-create — real democracy. I think they'll want to see much the same kind of participatory assemblies that they're likely to model themselves rolled out to supplement or (in cases of elite recalcitrance) replace the failing 'representative democracies' (aka elective dictatorships) that have overseen our lemming-like race to the eco-cliff-edge. And, of course, to modify or replace the failing authoritarian regimes that are doing the same, or in some cases (e.g. Putin's Russia) even worse.

And I think Parents For A Future will be deliberately broad-based, beyond ideology, beyond party politics and beyond sectional interests and 'identities'. They'll tend to be thoroughly inclusive, universalistic. They will have little patience for anything that divisively gets in the way of a truly adequate emergency-response to the mother of all crises, now threatening Mother Earth. They'll base everything in the call to do what it takes to save their kids (and *their* kids, and...). If this means (as it surely does) a fairly humungous redistribution of wealth, that outcome will be a result of the needs of our kids in the context of the long eco-emergency, not of any 'left-wing' ideology.[144]

Let me add a hope. I hope that they'll be brave enough to incorporate the agenda of 'transformative adaptation'[145] that I've touched on repeatedly in this book. I hope they'll be fuelled by rage — the righteous rage that springs from love for their most vulnerable. Rage that the world has left it too late to enjoy a smooth transition to a system that can last.[146] A smooth transition would have had to have begun in the 1960s, or at the very latest the 1980s. I hope they'll be honest and courageous enough to face the dreadful reality that things are going to get worse for our children for quite a long time to come even if we now truly do our best. This tragic truth is what makes it certain that the tide of eco-concern and action will keep rising for a long time to come. For the foreseeable, we can't prevent climate disasters.[147] They are coming; they are worsening. We can only seek to mitigate them in the true sense of that word. Which means adapting to what is here and what is coming in a manner that mitigates the force of the blow, shrinks as fast as possible the ongoing harm we are doing, and transforms our system to a better one: more local in its economics, more resilient, less materialistic, slower, more equal, more caring and relational, saner. (Those verbs and adjectives are transformative adaptation as I fashion it.) And, insofar as governments fail to deliver this, I hope that you, parents of the future, take it into your own hands, together, to change things in this way, in this direction. I hope that you won't wait around for 'them' to fix things, but that you'll get on with transforming your community, and what you can; because (y)our kids can't wait.

I say *you*, parents of the future — because this movement that is being proposed here cannot be led from the front by the likes of me.

Simply because I'm not a parent. I'd love to help advise, strategise and communicate, but, just as Fridays For Future had to be led by the kids themselves, so the movement that I'm now envisaging with you will have to be led by parents like you. Parenting the future is something that we can and must all do together; but it's *parents* who will need to lead on it.

That's why I say: it's over to you. If you have a better idea, good on you. If you don't have a better idea, then please consider this one, because it seems a very, very good one. But make it your own; the movement I'm proposing is nothing without *your authentic involvement*.

And be sure to remain authentic. That's been the secret of successful climate-communication in recent years, such as Greta's and XR's. Don't fantasise that you can make things just fine for your kids and theirs, if you do what is here proposed. Don't tell tall tales of stopping the eco-emergency dead in its track, nor fairy-stories of endless 'green growth'. Instead, harness the incredible, mostly still-untapped energy that comes from the anguish of understanding that we need to rise up not to make everything dandy, but to make the future more or less safely *survivable*... And, if we are lucky, and determined, one that future generations will once more be able to flourish in.

In the end, this is primal. This book is about what any mother knows in her bones, what any parent cannot deny, what any human who has ever felt a kinship with the next generation is. When we truly touch our essence, as parents of the future, as protectors of our vulnerable young, then we are unshakable... And then they, the powers that be, will hear our lion's roar.

Perhaps *you*, reader, can be a co-creator or even a leader of Parents For A Future.

Here's hoping.

If you rise to the call, then know that I eternally salute you, and, a trillion times more important, so do the children of the future.

References and Notes

Chapter 1

1   Ludwig Wittgenstein, *Philosophical Occasions 1912-1951*, ed. James Carl Klagge and Alfred Nordmann (Indianapolis: Hackett Publishing Company, 1993), 161.

2   Louis Dore, 'Society will collapse by 2040 due to catastrophic food shortages, says study', *The Independent*, 2020, https://www.independent.co.uk/environment/climate-change/society-will-collapse-by-2040-due-to-catastrophic-food-shortages-says-study-10336406.html. Many other sources could be cited here. Some are collected in Asher Moses's '"Collapse of civilisation the most likely outcome": top climate scientists', *Voice of Action*, 2020, https://voiceofaction.org/collapse-of-civilisation-is-the-most-likely-outcome-top-climate-scientists/. NB, Johan Rockström has since clarified that he thinks that the world population would 'only' drop to about four billion, in the event catastrophic climate breakdown: see, for example, the Scientists Warning website (https://www.scientistswarning.org/warnings/). See also Jem Bendell's website, in particular 'Climate science and collapse – warnings lost in the wind', 2020 (https:/jembendell.com/2020/06/15/climate-science-and-collapse-warnings-lost-in-the-wind/); and Pablo Servigne et al., 'Deep adaptation opens up a necessary conversation about the breakdown of civilisation', *openDemocracy*, 2020, https://www.opendemocracy.net/en/oureconomy/deep-adaptation-opens-necessary-conversation-about-breakdown-civilisation/.)

3   See Anna Leszkiewicz, 'TV's Climate Change Problem', *New States-man*, 2019, https://www.newstatesman.com/climate-change-television-big-little-lies-chernobyl-game-thrones. There have been various recent series which do feature climate decline, such as *Years and years* (2019), or which are allegorically relevant to the emergency, such as *Game of Thrones* (2011-2019), but none in which it has been central. A possible exception is the new *Snowpiercer* (2020–), but it is questionable whether this is 'significant', whereas the film on which the TV series is based most definitely was.

4   *The Road* (2009), directed by John Hillcoat, based on the 2006 post-apocalyptic novel by American writer Cormac McCarthy; *Mel-ancholia* (2011), directed by Lars von Trier; *Avatar* (2009), directed by James Cameron. I analyse *The Road* (alongside *Melancholia* and *Avatar*) in depth in my *A Film-Philosophy of Ecology and Enlightenment* (Andover: Routledge, 2019). There have been other blockbusters addressing the climate crisis head-on; most notably *The Day after Tomorrow* (2004) and *Geostorm* (2017). But I ignore these, as they are pap.

5   Amitav Ghosh, *The Great Derangement: Climate Change And The Unthinkable* (Chicago: University of Chicago Press, 2016).

6   I am referencing here the human-influenced weather of this emerging new geological era, increasingly called the Anthropocene.

7   Of course, this is not to deny that there are major, more or less novelistic, 'cli-fi' works dealing more or less successfully with the topic, far more than TV has done. Consider some of J.G. Ballard's work, or K.S. Robinson's and Saci Lloyd's. Plus, Margaret Atwood's *Oryx and Crake* (2003), Barbara Kingsolver's *Flight Behaviour* (2012), John Lanchester's *The Wall* (2019), Nancy Kress's *Nothing Human* (2003), Marcel Theroux's

Far North (2009), Paulo Bacigalupi's *The Windup Girl* (2009). These are deeply impressive books that have been hugely important to me. Though it is striking that they are mostly either quasi-epics or, to use a technical term, profoundly weird... It is not obvious or certain that they counteract Ghosh's claim that the novel as we have known it cannot encompass climate breakdown. In my opinion, the most impressive of all among the works just mentioned tend to be the least 'realist' (I'm thinking especially of Kress's and Theroux's).

8   See Pauline Loong, 'Coronavirus isn't a black swan event, says Asia-Analytica's Loong", *Bloomberg.com*, 2020, https://www.bloomberg.com/news/videos/2020-02-10/coronavirus-isn-t-a-black-swan-event-says-asia-analytica-s-loong-video; and Bernard Avishai, 'The pandemic isn't a black swan but a portent of a more fragile global system', *The New Yorker*, 2020, https://www.newyorker.com/news/daily-comment/the-pandemic-isnt-a-black-swan-but-a-portent-of-a-more-fragile-global-system.

9   See Emma Newburger, 'Wildlife habitat destruction and deforestation will cause more deadly pandemics like coronavirus, scientists warn, *CNBC*, 2020, https://www.cnbc.com/2020/05/09/coronavirus-wildlife-habitat-destruction-will-cause-more-pandemics.html.

10   See Kathrine Galagher, 'The connection between coronavirus and wildlife exploitation', *In Habitat*, 2020, https://inhabitat.com/the-connection-between-covid-19-and-wildlife-exploitation/; and Claire Anderson, 'Wuhan wet market horror laid bare as gruesome practice starts up AGAIN despite COVID-19', *Express*, 2020, https://www.express.co.uk/news/world/1266574/wuhan-wet-market-china-coronavirus-animal-covid-19-death-toll-latest-update.

11   See Joseph Norman, Yaneer Bar-Yam, and Nassim Nicholas Taleb, 'Systemic risk of pandemic via novel pathogens – coronavirus: a note',

New England Complex Systems Institute, 2020, https://necsi.edu/system-ic-risk-of-pandemic-via-novel-pathogens-coronavirus-a-note.

12 I draw out the huge importance of the vulnerability we have *experienced* around Covid for the possibility of our now acting adequately on climate etc., at Chapter 26, 'Theses on the coronavirus crisis', of my *Extinction Rebellion: Insights From The Inside*, ed. Samuel Alexander (Melbourne: Simplicity Institute Press, 2020). See also the Appendix to that book, on the same theme. It is our sense of actual logistical/physical vulnerability to climate chaos that could prove immeasurably more motiving than worthy talk of 'emissions' and 'parts per million'. Compare the way I framed this point on BBC TV's *Questiontime*, which can be found at https://www.youtube.com/watch?v=QK7DKiKh9_Q at the five minute mark.

13 For explication of this controversial claim, see my talk at Schumacher College on 'Eco-spirituality at the moment of last chance' at https://www.youtube.com/watch?v=4kbzI_jTGIk.

14 See Rupert Read, 'Imagining the world after COVID-19', *ABC*, 2020, https://www.abc.net.au/religion/rupert-read-imagining-a-world-after-coronavirus/12380676?utm_medium=social&utm_content=sf235359928&utm_campaign=abc_religion&utm_source=t.co&sf235359928=1.

15 For discussion, see David Wallace-Well's mini-essay: 'What Climate Alarm Has Already Achieved', *Intelligencer*, 2020, https://nymag.com/intelligencer/2020/08/what-climate-alarm-has-already-achieved.html.

16 I have in mind here the great book of Karl Polanyi's of that title (first published in 1944), and the way in which the transformation that we now need points in the *opposite* direction to that transformation

(basically, toward neoliberalism) that Polanyi described and deconstructed. To gain a sense of the great transformation that we now need, see Rupert Read, 'A discussion of Transformative Adaptation: a way forward for the 2020s', *YouTube*, 2020, https://www.youtube.com/watch?v=msvHevicz24&ab_channel=RupertRead.

17   Special issue of *Nordic Wittgenstein Review* on Post-Truth (2019), ed. Rupert Read and Timur Ucan, https://www.nordicwittgensteinreview.com/issue/view/245.

18   Rupert Read, 'What is New in Our Time: The Truth in "PostTruth"', *Nordic Wittgenstein Review* (2019): 81-96, DOI 10.15845/nwr.v8i0.3507, https://www.nordicwittgensteinreview.com/article/view/3507/4190.

19   Jean-Paul Sartre, *The Reprieve*, 1945 (London: Penguin Modern Classics, 2001).

20   For the kind of thing I have in mind here, see: Rupert Read and Deepak Rughani, 'Heartbreaking Genius of Over-Simplification', *Byline Times*, 2020, https://bylinetimes.com/2020/05/14/review-michael-moores-planet-of-the-humans-heartbreaking-genius-of-staggering-over-simplification/.

21   I'm proud to be among the select group of those who *have* taken them on, and won. See Ian Sinclair, 'No more climate cranks on our screens', *Morning Star,* 2020, https://morningstaronline.co.uk/article/no-more-climate-cranks-our-screens.

22   Damian Carrington, 'Insect numbers down 25% since 1990, global study finds', *The Guardian*, 2020, https://www.theguardian.com/environment/2020/apr/23/insect-numbers-down-25-since-1990-global-study-finds. 'Insectageddon' is contested by some. But again, the onus is on them; if there is even a risk that the picture is broadly accurate,

then it requires of us massive precautionary action. For insects are a non-negotiable part of most ecosystems, including many on which human food is based.

23   Brad Plumer and Nadja Popovich, 'The World Still Isn't Meeting Its Climate Goals', *New York Times*, 2018, https://www.nytimes.com/interactive/2018/12/07/climate/world-emissions-paris-goals-not-on-track.html. For detail on how Paris is inadequate, see also: Rupert Read, 'This civilisation is finished: So what is to be done?', *IFLAS Occasional Paper 3*, 2018, http://lifeworth.com/IFLAS_OP_3_rr_whatistobedone.pdf.

24   See Mark Lynas's 2007 book, *Six Degrees: Our Future on a Hotter Planet*, for some chapter and verse. See also Kerry Sheridan, 'Earth risks tipping into "hothouse" state: study', *phys.org*, 2018, https://phys.org/news/2018-08-earth-hothouse-state.html for a terrifying update, and John Gowdy, 'Our hunter-gatherer future: Climate change, agriculture and uncivilization', *Science Direct*, 2020, https://www.sciencedirect.com/science/article/pii/S0016328719303507 for discussion of why the only question is whether we allow our civilisation to be demolished, or intelligently transition to another civilisation.

25   'Climate change: Methane gas leaking from Antarctica seabed', *BBC*, 2020, https://www.bbc.co.uk/newsround/53503094 and Robert Hunziker, 'Thawing Arctic Permafrost, *Counter Punch*, 2020, https://www.counterpunch.org/2020/07/24/thawing-arctic-permafrost/.

26   See Nafeez Ahmed, 'Green economic growth is an article of "faith" devoid of scientific evidence', *Insurge Intelligence*, 2020, https://medium.com/insurge-intelligence/green-economic-growth-is-an-article-of-faith-devoid-of-scientific-evidence-5e63c4c0bb5e [https://t.co/NC2ADrZaPr?amp=1] for recent confirmation of this.

27 See Kevin Anderson, 'The hidden agenda: how veiled techno-utopias shore up the Paris Agreement', *kevinanderson.info*, 2016, http://kevinanderson.info/blog/the-hidden-agenda-how-veiled-techno-utopias-shore-up-the-paris-agreement/ for chapter and verse. For more about NETs, see Helena Paul and Rupert Read, 'Geoengineering as a Response to the Climate Crisis: Right Road or Disastrous Diversion?', *Face Up to Climate Reality: Honesty, Disaster and Hope*, ed. John Foster (London: Green House, 2019), 109-131, https://www.greenhousethinktank.org/uploads/4/8/3/2/48324387/fucr_foster_chapter_6_updated.pdf (my co-authored critique of all such 'geoengineering' as unprecautious). Cf. also Gowdy, 'Our hunter-gatherer future'.

28 Rupert Read, 'Climate Change is a White Swan', *The Ecologist*, 2017, https://theecologist.org/2017/feb/23/climate-change-white-swan.

29 For detail, see Wallace-Wells, 'What Climate Alarm Has Already Achieved'.

30 For some instances of where reality already appears to be exceeding the 'worst case scenarios' that scientists had postulated, this report is well worth reading: David Spratt and Ian Dunlop, *What Lies Beneath: The Understatement of Climate Risk* (Melbourne: Break Through: National Centre for Climate Restoration, 2018), https://52a87f3e-7945-4bb1-abbf-9aa66cd4e93e.filesusr.com/ugd/148cb0_a0d7c18a1b-f64e698a9c8c8f18a42889.pdf; and it's worth noting this more recent chastening example: Scott Snowden, 'Greenland's Massive Ice Melt Wasn't Supposed to Happen Until 2070', *Forbes*, 2019, https://www.forbes.com/sites/scottsnowden/2019/08/16/greenlands-massive-ice-melt-wasnt-supposed-to-happen-until-2070/. Others could be cited. For a cautionary note on why we shouldn't be too trustful in models, not for the reason that 'climate sceptics' assert (that they are too negative), but for the opposite reason (that they may lull us into a full sense of

complacency), see Nassim Nicholas Taleb and Rupert Read et. al., 'Climate models and precautionary measures', *The Black Swan Report*, 2015, https://www.blackswanreport.com/blog/2015/05/our-statement-on-climate-models/. See also, Wolfgang Knorr, 'The Climate Crisis Demands New Ways of Thinking from Climate Scientists', *Resilience*, 2020, https://www.resilience.org/stories/2020-08-04/the-climate-crisis-demands-new-ways-of-thinking-from-climate-scientists/.

31 See Whit Gibbons, 'The Legend of the Boiling Frog is Just a Legend', *Savannah River Ecology Lab Archive*, University of Georgia, 2007, http://archive-srel.uga.edu/outreach/ecoviews/ecoview071223.htm.

32 In the UK, principally via the Green Deal scheme, which is very generous: 'Green Deal: energy saving for your home', https://www.gov.uk/green-deal-energy-saving-measures.

Chapter 2
33 Immanuel Kant, *Political Writings*, ed. H.S. Reiss (Cambridge: Cambridge University Press, 1970), 50.

34 See Fred Pearce, 'Global Extinction Rates: Why Do Estimates Vary So Wildy?', *Yale Environment 360*, 2015, https://e360.yale.edu/features/global_extinction_rates_why_do_estimates_vary_so_wildly. As explained in Chapter 3, the very fact that we don't know how many species there are exposes us (and them) to elevated risk. Actual species extinction rates might be orders of magnitude lower than this (cf. e.g. 'Current Extinction Rate 10 Times Worse Than Previously Thought', *IFL Science!*, https://www.iflscience.com/plants-and-animals/current-extinction-rate-10-times-worse-previously-thought/). But it would be unacceptably rash to assume that they are. We must actively consider worst-case scenarios / 'fat tails', when it comes to extinction, which is forever.

35 David Archer, et. al., 'Atmospheric Lifetime of Fossil Fuel Carbon Dioxide', *Annual Review of Earth and Planetary Sciences*, vol. 3, no. 7 (2009): 117-34, https://www.annualreviews.org/doi/abs/10.1146/annurev.earth.031208.100206.

36 The UK Government led the world by legislating for a zero target for carbon, by 2050, in the wake of the April 2019 Extinction Rebellion, which had demanded exactly that, but by 2025. (Actually, the target is for *net*-zero carbon, which is considerably more problematic. See 'A Reliance on Negative Emissions Technologies is Locking in Carbon Addiction', *Geoengineering Monitor*, 2016, http://www.geoengineeringmonitor.org/2016/10/a-reliance-on-negative-emissions-technologies-is-locking-in-carbon-addiction/ and Kevin Anderson, 'Brief Response to the UK Government's "Net-Zero" Proposal, *Resilience*, 2019, https://www.resilience.org/stories/2019-06-18/brief-response-to-the-uk-governments-net-zero-proposal/.)

37 This would need to exceed considerably what demographers actually predict: *World Population Prospects*, United Nations Department of Economic and Social Affairs Population Division, World Population Prospects, 'Demographers estimate the population could almost stop growing in about 2100', 2019, https://population.un.org/wpp/Publications/Files/WPP2019_Highlights.pdf.
Of course, as noted in the main text, there *is* a credible reason, to which I will return, why the population of the Earth might drastically decrease: collapse. But that of course is exactly the scenario that the thinking of this essay is designed to try to *prevent*.

38 My argument in this chapter is without doubt in considerable tension with certain arguments in Lee Edelman's No Future: *Queer Theory and the Death Drive* (Durham: Duke University, 2004). But on this key point, which is Edelman's very starting point, there is *no*

disagreement between us. Any ruthlessly heteronormative account of futurity is not acceptable. (And, more basically, not 'necessary'. The close association between heterosexuality and the parental role has, thankfully, been broken.)

39 *Children of Men* (2006), dir. Alfonso Cuaron, a dystopian action thriller, loosely based on the PD James novel of the same name (1992).

40 Of course, strictly speaking it is misleadingly anthropocentric to regard ourselves as 'the dominant species'. We have vast power, but, crucially, power that is not properly in our control; and, in any case, the coronavirus has been a powerful reminder of how other Earthly beings (including the very smallest) can humble us.

41 For some detail to support this claim, see my critical consideration of another example of tech-fixery, robotization: Rupert Read, 'The Rise of the Robot: Dispelling the myth', *The Ecologist*, 2016, https://theecologist.org/2016/dec/13/rise-robot-dispelling-myth.

42 I make the argument that this is what dangerous anthropogenic climate change is in Rupert Read and Phil Hutchinson, 'Wittgenstein and Pragmatism', *Cambridge Companion to Pragmatism*, ed. Alan Malachowski (Cambridge: Cambridge University Press, 2013). Cf. also Timothy Morton's broadly-allied conception of climate change as a 'hyper-object', as discussed in Andrea Gibbons, 'Timothy Morton: Hyperobjects, Climate Change (and Trump)', *writingcities.com*, 2016, http://writingcities.com/2016/11/10/timothy-morton-hyperobjects-climate-change-trump/.

43 Chris D'Angelo, 'Chasing the Methane Dragon That Lurks In The Deep Sea', *Huffington Post*, 2019, https://www.huffingtonpost.co.uk/entry/methane-hydrate-atlantic-samantha-joye_n_5d681737e4b-

0488c0d117841. See also Facing Future, 'Dr. Peter Wadhams: Arctic Research & the Methane Risk', *YouTube*, 2019, https://www.youtube.com/watch?v=D3L0R6LzEUE for an interview with a leading scientific expert on the methane threat; and Environmental Coffeehouse, 'Dr. Peter Wadhams speaks out. Artic Sea Ice, Methane and more', *YouTube*, 2020, https://www.youtube.com/watch?v=KxDg3pgbW9g&feature=youtu.be for a provocative update. The likelihood of any methane depth-charge is hotly contested. It may well be extremely unlikely. But once more, a very high threshold of confidence indeed would be needed, to dismiss it altogether — given the vast/infinite downside, if a 'methane bomb' goes off. The latest indications at going to press are not unworrying: 'Methane forecasts', Atmosphere Monitoring Service, https://atmosphere.copernicus.eu/charts/cams/methane-forecasts?facets=undefined&time=2020102900,3,2020102903&projection=classical_arctic&layer_name=composition_ch4_surface.

44 Based on a line from William Shakespeare's 'Sonnet 116': 'That looks on tempests and is never shaken' (line 6).

45 This exercise is inspired by a very similar exercise in chapter 4 of Roman Krznaric's book, *The Good Ancestor: How to Think Long Term in a Short-term World* (London: WH Allen, 2020).

46 This figure is drawn from John Passmore's Man's *Responsibility for Nature: Ecological Problems and Western Traditions* (London: Duckworth, 1974).

47 See Kevin Hester, 'This Civilisation is Finished. Rupert Read, Paul Ehrlich and Jem Bendell', *kevinhester.live*, 2018, https://kevinhester.live/2018/12/28/this-civilisation-is-finished-ruppert-reid-paul-ehrlich-and-jem-bendell/ for succinct back-up for the point made here, and David Fleming's 'Lean Guide to Nuclear Energy', *Nuclear Energy*, http://www.theleaneconomyconnection.net/nuclear/index.html) for some

more detailed back-up. The claim that nuclear is collapse-unsafe is (if they ever deign to consider it, which they do too rarely) contested by advocates of nuclear. But the burden of proof is very much upon them. A very high threshold of confidence would be needed for us to judge this ultra-complex technology, with high potential downsides, as safe *in extremis*. For some thinking as to why that threshold is unlikely to be reached, see the discussions of Fukushima/nuclear found at more than one point in Nassim Nicholas Taleb's *Antifragile: Things That Gain from Disorder* (New York: Penguin Random Hosue, 2012).

48 This portends the Precautionary Principle (See Rupert Read, 'The Precautionary Principle', *rupertread.net*, https://rupertread.net/precautionary-principle), which I detail in Chapter 4.

49 This example comes from my teacher the late Derek Parfit's *Reasons and Persons* (Oxford: Oxford University Press, 1987), 357.

50 Cf. again Krznaric's book-length treatment, *The Good Ancestor*.

51 For more detailed back-up of this claim, see the collapsologists' impressive discussion here: Servigne et. al., 'Deep adaptation opens up a necessary conversation'.

52 See my account, 'Imagining the World after Covid-19', of how the coronavirus is a product of growth-driven habitat-destruction, of mal-treatment of animals, and of out-of-control air travel.

53 J.R.R. Tolkien, *The Return of the King* (London: George Allen & Unwin, 1955), 65.

Chatper 3

54 *Excalibur* (1981), dir. John Boorman; these words are spoken to King Arthur (but my implication is that in our more democratic age they should be heard as applying to us all).

55 What I am doing in this chapter, in philosophical terms, is radicalizing further the subtlest and most serious form of prudential anthropocentrism: 'the convergence hypothesis.' That truly placing the deep human future centrally converges with placing ecosystems centrally. In my terms, that real anthropocentrism *collapses into* ecocentrism. Here is Bryan Norton's formulation, from his *Sustainability: A Philosophy of Adaptive Ecosystem Management* (Chicago: University of Chicago Press, 2005): 'The Convergence Hypothesis ... leads us to embrace a useful formulation developed by the deep ecologists, who insist that we are not really separate from nature; our skin is but a permeable membrane ... [I]nsults to my immediate environment are insults to the broader self I embrace... In a nondualistic world, where humans are never severed from nature ... the environment ... is as much a part of us as we are a part of it' (37).

56 Emily Chung, '60% of world's wildlife has been wiped out since 1970', *CBC*, 2018, https://www.cbc.ca/news/technology/living-plant-wwf-2018-1.4882819.

57 Joydeep Gupta, 'Scientists Warm of a Looming Mass Extinction of Species, *The Wire*, 2019, https://thewire.in/environment/global-bio-diversity-extinction-scientists-climate-change. See also University of Exeter, 'Biodiversity loss raises risk of "extinction cascades"', *phys.org*, https://phys.org/news/2018-02-biodiversity-loss-extinction-cascades.html for the way this could play out more cumulatively than one might naively expect.

58   Actually, I doubt, on philosophical grounds, that this is so: I would argue, drawing on Wittgenstein, Merleau-Ponty and Hubert Dreyfus among others, that it doesn't mean anything to suppose that humanity and our capacity for wisdom could leave embodiedness behind. But I don't need to press that point here; for the precautionary reason given in the text, which is enough.

59   See e.g. Ian Thompson, et. al., *Forest Resilience, Biodiversity, and Climate Change*, Secretariat of the Convention on Biological Diversity, Montreal, Technical Series, no. 43, 2009, https://www.cbd.int/doc/publications/cbd-ts-43-en.pdf. More generally, see E.O. Wilson, *Half-Earth: Our Planet's Fight for Life* (New York: Norton, 2016).

60   Camilo Mora, et. al., 'How Many Species Are There on Earth and in the Ocean?', *PLoS Biology*, vol. 9, no. 8 (2011), https://journals.plos.org/plosbiology/article?id=10.1371/journal.pbio.1001127: Scientists from the Census of Marine Life estimated there were about 8.7 million (give or take 1.3 million) species on Earth with 86 per cent of all species on land and 91 per cent of those in the seas have yet to be discovered. *But* there is actually *considerably greater* uncertainty about such estimates than this study admitted: as laid out in Chapter 3 of E.O. Wilson's *Half-Earth*.

61   Formulated by scientist, environmentalist and futurist James Lovelock (along with microbiologist Lynn Margulis) in the 1960s. The Gaia hypothesis proposes that living and non-living parts of the Earth form a complex interacting system that can be thought of as a single organism.

62   Unlike many – e.g. the Pope in his beautiful work, Laudato Si (See Fr. Robert Barron, 'Bishop Barron on Pope Francis' Encyclical "Laudato Si", *Word on Fire*, 2015, https://laudatosi.com/watch). I do not take the liberty of assuming that nature/life is intrinsically valuable. I do not assume that you grant that, but rather seek to lead you towards

a conclusion roughly along those lines starting from a more limited basis that you surely *will* grant, because it is widely shared and unobjectionable (even from the objectionable...). If proof be needed of the virtue of Pope Francis's eco-encyclical, it can be conveniently found in in the fact of this attack upon it, hosted by the climate-denialist, *The Global Warming Policy Forum*: Mark Lynas, Ted Nordhaus and Michael Shellenberger, 'A Pope Against Progress', *The Global Warming Policy Forum*, 2015, http://www.thegwpf.com/a-pope-against-progress/.

63 Rupert Read, 'Are Some Risks Just Too Big To Take?', *University of East Anglia*, 2017, https://www.uea.ac.uk/research/explore-uea-research/are-some-risks-just-too-big-to-take.

64 Cf. Rupert Read, 'APPG Briefings on the Precautionary Principle (Climate Change and Animal Welfare)', *APPG*, 2018, https://agroecology-appg.org/ourwork/appg-briefings-on-the-precautionary-principle-climate-change-and-animal-welfare/.

65 See 'How sensitive is our climate', *Skeptical Science*, https://skepticalscience.com/climate-sensitivity-advanced.htm. See also 'Climate sensitivity on the rise?', *Met Office*, 2018, https://www.metoffice.gov.uk/research/news/2018/climate-sensitivity; and David Carlin, 'Disaster or Relief: Why Climate Sensitivity Matters', *Forbes*, 2020, https://www.forbes.com/sites/davidcarlin/2020/08/10/disaster-or-relief-why-climate-sensitivity-matters/#2834061570c1.

66 For the essential point *vis-a-vis* how one side-steps and marginalizes 'climate-scepticism' through the kind of precautionary considerations I'm marshalling here, see my short joint piece with Taleb et. al., 'Climate models and precautionary measures'.

67   See Nick Breeze, 'It's nonlinearity – stupid!', *Ecologist*, 2019, https://t.co/TdqI92XB0g for leading German climate expert Hans Schnellnhuber's view on the credible risk of civilisational-ending giga-death catastrophe; and Gaia Vince, 'The heat is on over the climate crisis. Only radical measures will work', *The Observer*, 2019, https://www.theguardian.com/environment/2019/may/18/climate-crisis-heat-is-on-global-heating-four-degrees-2100-change-way-we-live for the doyen of planetary boundaries, Johan Rockstrom, on the same.

68   The clearest scientific demonstration of this is Johan Rockstrom's 'Planetary boundaries research', *Stockholm University: Stockholm Resilience Centre*, https://stockholmresilience.org/research/planetary-boundaries.html.

69   See Damian Carrington, 'Coronavirus: "Nature is sending us a message", says UN environment chief", *The Guardian*, 2020, https://www.theguardian.com/world/2020/mar/25/coronavirus-nature-is-sending-us-a-message-says-un-environment-chief?CMP=Share_iOSApp_Other.

70   Contrary to what is often supposed, the main threat facing biodiversity (life) *to date* is not climate-damage, but habitat-destruction (especially via agriculture). See e.g. 'Media Release: Nature's Dangerous Decline "Unprecedented"; Species Extinction Rates "Accelerating"', https://ipbes.net/news/Media-Release-Global-Assessment; Robert Watson, 'Loss of biodiversity is just as catastrophic as climate change', *The Guardian*, 2019, https://www.theguardian.com/commentisfree/2019/may/06/biodiversity-climate-change-mass-extinctions; and Helen Santoro, 'The World is Failing to Stop Extinctions. These Scientists Have a Plan to Help', *Huffington Post*, 2020, https://www.huffingtonpost.co.uk/entry/scientists-save-endangered-species-extinction_n_5f32a91fc5b6fc009a5dbded?ri18n=true.

71 Joanna Macy, *World as Lover, World as Self* (Berkeley: Parallax Press, 1991).

72 I was taught this exercise by Joanna Macy.

73 Edward O. Wilson, *Biophilia* (Cambridge: Harvard University Press, 1984).

74 The moment, perhaps already in the past, when known oil reserves start shrinking. See Thom Hartmann, *The Last Hours of Ancient Sunlight* (New York: Mythical Books, 1998). Or Rachel Dobbs, 'What is peak oil? And what will happen to the industry if we reach it?', *Verdict*, 2018, https://www.verdict.co.uk/peak-oil/ for a more 'mainstream' perspective.

75 'The Ecological Footprint shows that people are using the capacity of 1.5 Earths – but how can this be when there is only one Earth?', https://wwf.panda.org/knowledge_hub/all_publications/living_planet_report_timeline/lpr_2012/demands_on_our_planet/overshoot/

76 Derrick Jensen, *Endgame Vol.2: Resistance* (New York: Seven Stories Press, 2006), 578.

77 Yes, I'm aware that kids also sometimes pull the wings off flies and much worse! But most children, most of the time, easily slip into a closer kinship with nature and the wild, including (crucially) imaginatively, than most of us 'disenchanted' adults. Much more so, in cultures (such as most Indigenous and some peasant cultures) which take as given, and educate kids into, the kind of attitude of respect for our common home that this chapter is seeking to recommend we adopt.

78 Ursula K Le Guin (1929-2018). Author of twenty-one novels, eleven volumes of short stories, four collections of essays, twelve children's

books, six volumes of poetry and four of translation. The notion of 'propertartarianism' is coined in her classic of utopia-in-progress/dystopia-in-stasis, *The Dispossessed* (New York: Harper & Row, 1974).

79 Edward O. Wilson, *Biodiversity* (Washington: National Academies Press, 1988). See, in particular, Part 8: 'Restoration Ecology: Can We Recover Lost Ground?', 311-54.

Chapter 4

80 Founded by Swedish schoolgirl Greta Thunberg in August 2018, Strike For Climate (also known as Fridays For Future, Youth for Climate, Climate Strike or Youth Strike for Climate) has grown to an international movement where children, where or not doing so is 'permitted', take time out from their education to demonstrate and campaign.

81 Bruno Latour, *Down to Earth* (Oxford: Polity Press 2018), 13. Originally published in French, 2017.

82 Patrick Geddes, *Cities in Evolution* (London: Williams & Norgate, 1915). Although the exact phrase does not appear in the book, the idea is clearly evident in his words.

83 See Molly Larkin, 'What is the 7th Generation Principle and why do you need to know about it?', *mollylarkin.com*, https://mollylarkin.com/what-is-the-7th-generation-principle-and-why-do-you-need-to-know-about-it-3/; and '7th Generation Principle', *Seven Generations International Foundation*, http://7genfoundation.org/7th-generation/.

84 For detail, see Krznaric, *The Good Ancestor*, Chapter 6 'Cathedral Thinking: The Art of Planning into the Distant Future'.

85　In 2012 the Communist Party of China officially adopted an 'ecological civilization' as part of its constitution.

86　Itamar Zohar, 'Report: China Bans Avatar From 1,600 Cinemas Due to Fear of Popular Revolt', *Haaretz*, 2010, https://www.haaretz.com/1.5049164.

87　Initiated in 2019, 150 French citizens, chosen at random, make up the Citizens' Assembly on Climate. See Yannick Ondoa, 'President Macron keeps 146 of the 149 proposals of the Citizens Convention for Ecology', *Yannick Ondoa: Medium*, 2020, https://medium.com/@yannick-ondoa91200/president-macron-keeps-146-of-the-149-proposals-of-the-citizens-convention-for-ecology-adc449dc7ac2 on its recommendations.

88　See 'Climate: the Citizens' Convention makes 50 proposals to create "a different model"', *L'Express*, 2020, https://www.lexpress.fr/actualite/societe/environnement/climat-la-convention-citoyenne-fait-50-propositions-pour-creer-un-modele-different_2123437.html.

89　A lot of hope is being invested at present in the idea of a Green New Deal. This is a promising sign. It will all probably come to nothing unless the idea is understood as being for the wider benefit of society (as opposed to the playing out of a left-wing ideology). More important still: I use the word 'real' because much that is called a 'Green New Deal' is little more than greenwashed industrial growthism. A real Green New Deal would step beyond the defunct aim of economic growth. For some discussion, see Rupert Read and Frank Scavelli, 'Sanders' Green New Deal: A Realistic Response to the Emergency That Will Define Our Lifetimes', *Common Dreams*, 2019, https://www.commondreams.org/views/2019/11/25/sanders-green-new-deal-realistic-response-emergency-will-define-our-lifetimes; and 'Plan B vs. Plan C', *Bright Green*, 2011, https://bright-green.org/2011/10/31/plan-b-vs-plan-c/. Cf. also Richard Murphy, 'Growth, MMT and the Green

New Deal', *Tax Research UK*, 2019, https://www.taxresearch.org.uk/ Blog/2019/05/21/growth-mmt-and-the-green-new-deal/; and 'Pollution caps and modern monetary theory', *Tax Research UK*, 2019, https:// www.taxresearch.org.uk/Blog/2019/02/04/pollution-caps-and-modern-monetary-theory/. As so often, Greta Thunberg has a very sharp perspective on this matter (listen to 'Greta Thunberg: Humanity has not yet failed' on Sverige Radio's podcast *Summer & Winter in P1*, 2020, https://sverigesradio.se/sida/avsnitt/1535269?programid=2071), helping to puncture the wishful thinking that characterises too much agitation for a Green New Deal. She remarks: 'If rich countries like Sweden and the UK are to fulfil their commitments in the Paris Agreement, they need to reduce their total national emissions of CO2 by twelve to fifteen per cent every year, starting now [Here, she draws upon this recent study: 'UK & Sweden's carbon targets half what is needed for 1.5C', *Tyndall Centre for Climate Change Research*, 2020, https://tyndall. ac.uk/news/uk-sweden's-carbon-targets-half-what-needed-15c]. Of course, there's no green recovery plan or deal in the world that alone would be able to achieve such emissions cuts. And that's why *the whole Green Deal debate ironically risks doing more harm than good, as it sends a signal that the changes needed are possible within today's societies.* As if we could solve a crisis without treating it as a crisis' (emphasis added).

90   By this phrase, I mean most of all to question the credentials of most so-called 'bioenergy', including crucially large-scale biofuels. See on this the (crucial) work of *Biofuel Watch* (https://www.biofuelwatch.org.uk). In any case, Citizens' Assemblies would clearly need to think through the question of what actually deserves to be called renewable energy. Cf. on this: Read and Rughani, 'Heartbreaking Genius of Staggering Over-Simplification'.

91   See Flora Southey, 'Food rationing: UK urged to adopt health-based scheme, "not ad hoc led by retailers"', *Food Navigator*, 2020, https:// www.foodnavigator.com/Article/2020/03/23/Coronavirus-UK-urged-to-

adopt-health-based-food-rationing-scheme-not-ad-hoc-led-by-retailers
for the Covid precedent for such a call.

92   To any ultra-'PC' readers who want to deny that there is an overpop-
ulation issue at all — even once nuanced (and this nuance, clear in my
text above, is critical) by stressing that the more 'development', the more
'wealth', the more damaging the impact of any human population (such
that the most damaging are populations of rich countries with high footprints) — let me press this question: if the demographic projections we
are being offered (of billions more mouths to feed, in years to come) play
out, and human populations continue to rise, where are all the animals
supposed to go? Where is the restored / rewilded land supposed to be?
And (adding in a precautionary dimension) where is the safety-margin?
Now that we are in terra incognita population-wise, every additional human, and every additional bit of economic activity, takes us further from
safety and further into risk of black/grey swan events. In answer to these
questions, it isn't enough to say sensible vanilla things like 'more of us
should adopt a vegan diet'. These questions that I've just asked make
clear the negative trade-offs that are bound to result from an even higher
human population / economic impact upon the planet. John Ruskin said,
beautifully, that the only true wealth is life. At the end of the day, excessive amounts of human life mean less of other life, to the detriment of
that life and (as I argued in Chapter 3) to that of humans too. That is why,
in Chapter 2, I suggested that one important way to help your kids is...
not to have too many of them. (See Caroline Mortimer, 'Having children
is one of the most destructive things you can to do [sic] the environment,
says researchers', *Independent*, 2017, https://www.independent.co.uk/
environment/children-carbon-footprint-climate-change-damage-having-
kids-research-a7837961.html for factual back-up for this thought.)

93   To understand better the conception of transformative adaptation
that I have developed, see my open Google Document, 'An introduction to

transformative adaptation', https://docs.google.com/document/d/1lAW-JxPFbV7IuShx2ShSIzN1yORL-v5_BgAbUtcAqTNI/edit. See also Rupert Read and Samuel Alexander, *Extinction Rebellion: Insights from the Inside* (Melbourne: Simplicity Institute, 2020), especially pages 44-5 [This book is available for free download at https://249897.e-junkie.com/product/1668648]. And see *Facing Up to Climate Reality – Honesty, Disaster and Hope*, ed. John Foster (Lancaster: Green House Think Tank, 2019).

94 See e.g. Nafeez Ahmed, 'Theoretical Physicists Say 90% Chance of Societal Collapse Within Several Decades', *Vice*, 2020, https://www.vice.com/en_us/article/akzn5a/theoretical-physicists-say-90-chance-of-societal-collapse-within-several-decades.

95 This is the argument powerfully made by the collapsologists — see e.g. Servigne, et. al., 'Deep Adaptation'.

96 Rupert Read and Samuel Alexander, *This Civilisation Is Finished: Conversations at the End of Empire*, ed. Samuel Alexander (Melbourne: Simplicity Institute, 2019). This book discusses collapse (and its still-conceivable avoidance) in some detail.

97 Jem Bendell, 'Deep Adaptation: A Map For Navigating Climate Tragedy', *jembendell.com*, 2018, https://jembendell.com/2019/05/15/deep-adaptation-versions/. NB: As should be very clear, my view, unlike Bendell himself, is that we do not know that collapsing is bound to happen. We need to prepare for it as an eventuality to hedge against, not as a supposed 'certainty'. See Rupert Read, 'After the IPCC report, #climatereality', *Rupert Read: Medium*, 2018, https://medium.com/@rupertread_80924/after-the-ipcc-report-climatereality-5b3e2ae43697 for my (constructive) critique of Bendell's stance. See also our forthcoming co-edited book, *Deep Adaptation: Navigating Climate Chaos* (Oxford: Polity Press, 2021); this book will also be a report to the Club of Rome.

98 Dr Angela N. Baldwin and Sony Salzman, 'Yes, COVID-19 is mutating, here's what you need to know', *abc News*, 2020, https://abcnews.go.com/Health/covid-19-mutating/story?id=70535183.ry?id=70535183.

99 See Read, 'An introduction to transformative adaptation' for a full explication of the form of adaptation I most recommend: *transformative*. See also Read, 'A discussion of Transformative Adaptation' for a video discussion. As I present it, transformative adaptation includes deep adaptation, but has a wider and somewhat more hopeful agenda. (Because it views deep adaptation as an insurance policy, rather than viewing collapse as an inevitability as Jem Bendell himself does.)

100 Edmund Burke (1729 – 97). Irish statesman and philosopher. Served as an MP between 1766 and 1794 in Great Britain with the Whig Party.

101 Thomas Paine (1737 – 1809). Norfolk-born political philosopher and writer who supported revolutionary causes in America and Europe.

102 Charles Foster, *Being A Beast* (London: Profile Books, 2016).

103 Rupert Read, *Guardians of the Future – A Constitutional Case for representing and protecting Future People* (Dorset: Greenhouse, 2012), https://www.greenhousethinktank.org/uploads/4/8/3/2/48324387/guardians_inside_final.pdf.

104 From the UN Commission on Environment and Development Report published in 1987 as Our Common Future. For the incapacity of 'sustainable development', see John Foster, *After Sustainability: Denial, Hope, Retrieval* (Abingdon: Routledge, 2014). See also Helena Norberg-Hodge, *Ancient Futures: Learning from Ladakh* (Oakland: Sierra Club Books, 1991).

105 Joanna Macy, *Widening Circles: A Memoir* (Gabriola Island: New Catalyst Books, 2007).

106 Nassim Nicholas Taleb and Rupert Read, et. al., 'The Precautionary Principle (with Application to the Genetic Modification of Organisms)' *Extreme Risk Initiative – NYU School of Engineering Working Paper Series*, 2014, https://arxiv.org/abs/1410.5787.

107 See e.g. Morgan McFall-Johnsen, 'Greenland's ice is melting at the rate scientists thought would be our worst-case scenario in 2070', *Business Insider*, 2019, https://www.businessinsider.com.au/greenland-ice-melting-is-2070-worst-case-2019-8; and Dahr Jamail, 'As Antarctic Melting Accelerates Worst Case Scenarios May Come True', *Truthout*, 2018, https://truthout.org/articles/as-antarctic-melting-accelerates-worst-case-scenarios-may-come-true/.

108 See Read, 'The Precautionary Principle', https://rupertread.net/precautionary-principle.

109 See Rupert Read and Atus Mariqueo-Russell, 'Fully automated luxury barbarism', *Radical Philosophy*, vol. 2, no. 6 (2019), https://www.radicalphilosophy.com/reviews/individual-reviews/fully-automated-luxury-barbarism.

110 I am referring again here to those 'post-humanists' who think they can survive death. In a new literalistic version of an old religious fantasy, a version that takes shape via frozen skulls or via plans (sic) to be 'uploaded to the cloud'.

111 See Knorr, 'The Climate Crisis Demands New Ways of Thinking', on this point.

112 Nassim Nicholas Taleb, *The Black Swan: The Impact of the Highly Improbable* (New York: Penguin Random House, 2007). See Norman, Bar-Yam, and Taleb, 'Systemic risk of pandemic via novel pathogens'; and Rupert Read, 'What would a precautionary approach to the coronavirus look like?', *Rupert J Read: Medium*, 2020, https://medium.com/@rupertjread/what-would-a-precautionary-approach-to-the-coronavirus-look-like-155626f7c2bd.

Chapter 5

113 'Transcript: President Obama At Sandy Hook Prayer Vigil', *NPR*, 2012, https://www.npr.org/2012/12/16/167412995/transcript-president-obama-at-sandy-hook-prayer-vigil.

114 For what is meant by this claim, see Kristen Steele, 'Disaster Localization: A Constructive Response to Climate Chaos', *Resilience*, 2019, https://www.resilience.org/stories/2019-10-29/disaster-localization-a-constructive-response-to-climate-chaos/.

115 See e.g. Franziska Gaupp, et. al., 'Increasing risks of multiple breadbasket failure under 1.5 and 2° C global warming', *Agricultural Systems*, vol. 175 (2019), 34-45, https://www.sciencedirect.com/science/article/abs/pii/S0308521X18307674. This topic is not as well-researched as it should be, which is worrying in itself. Nevertheless, an increasingly clear worrying picture is emerging from the research we do have. Cf. Jem Bendell, 'Notes on Hunger and Collapse', *jembendell.com*, 2019, https://jembendell.com/2019/03/28/notes-on-hunger-and-collapse/ for a provocative overview. See 'Climate change to steepen food prices', *phys.org*, 2020, https://phys.org/news/2020-08-climate-steepen-food-prices.html for the worrying scenarios emerging in Africa. See also '2015 – Food insecurity and Climate Change Map', *United Nations World Food Programme*, 2015, https://www.wfp.org/publications/2015-food-in-

security-and-climate-change-map, though this is now out of date; in the five years since it was produced, as our climate has started seemingly spinning beyond our control, food scenarios have mostly worsened. And see the resources collected at Pamela Candea, *Food Impacts due to Climate Change*, 2019, https://futurescanning.files.wordpress.com/2019/04/food-impacts-due-to-climate-change-april-19.pdf. Covid-19 was briefly a somewhat-chastening dry-run for an eco-driven food shortage that could easily affect many more countries than the usual suspects in the Global South. Consider this piece about the views of the leading UK thinker on this topic, on the danger of food shortages in the UK: Prof. Tim Lang quoted in 'Food experts call for rationing to see Britain through pandemic', *Morning Star*, 2020, https://morningstaronline.co.uk/article/b/food-experts-call-rationing-see-britain-through-pandemic. (See also my evocation of the worry, live on the BBC's 'Question Time' programme: Extinction Rebellion, 'BBC Question Time | Dr. Rupert Read | Extinction Rebellion', *YouTube*, 2019, https://www.youtube.com/watch?v=QK7DKiKh9_Q . Go 5 minutes in.) At time of writing, autumn 2020, the worst — weather-ruined — grain harvests in more than a generation are being harvested/abandoned in parts of China, the USA, the EU and the UK, among other places. This is not an issue for the future only. This is already here now.

116 See Rupert Read, 'Negotiating the Space Between Apocalypse and Victory', *Byline Times*, 2020, https://bylinetimes.com/2020/06/12/negotiating-the-space-between-apocalypse-and-victory/.

117 Vladimir Lenin, *What is To Be done? Burning Questions of our Movement*, 1902 (London: Wellred Books, 2019).

118 J.R.R. Tolkien, *The Fellowship of the Ring* (London: George Allen & Unwin, 1954).

119 See Derrick Jensen, 'Forget Shorter Showers', *derrickjensen.org*, 2012, https://derrickjensen.org/2009/07/forget-shorter-showers/.

120 See Andrew Dobson, *Green Political Thought* (New York: Routledge, 1990) for explication. See also Rupert Read, 'How ecologism is the true heir of both socialism and conservatism', *LSE*, 2013, https://blogs.lse.ac.uk/politicsandpolicy/how-ecologism-is-the-true-heir-of-both-socialism-and-conservatism/.

121 For a compelling vision of this, see Chris Smaje's new book, *A Small Farm Future* (London: Chelsea Green, 2020).

122 See 'Climate and Ecological Emergency Bill', *CEE Bill Alliance,* https://www.ceebill.uk.

123 See Rupert Read, 'The coronavirus letter you've just been sent by Johnson is a lie', *YouTube*, 2020, https://www.youtube.com/watch?v=aKTwB-bge4lQ for back-up of this claim.

124 See 'How coronavirus has led...', *New Statesman*, 2020, https://www.newstatesman.com/international/2020/04/how-coronavirus-has-led-return-precautionary-principle.

125 See the exposé that Nafeez Ahmed and I did of the British Government's failure to follow the Precautionary Principle in *Byline Times*: 'Documents Reveal Government and NERVTAG Breached Own Scientific Risk Assessment Guidance', 2020, https://bylinetimes.com/2020/04/23/the-coronavirus-crisis-documents-reveal-government-and-nervtag-breached-own-scientific-risk-assessment-guidance/.

126 See Rupert Read, '24 Theses on Corona', *Rupert J Read: Medium*, 2020, https://medium.com/@rupertjread/24-theses-on-corona-748689919859.

127 See 'After coronavirus, focus on the climate emergency: Letters', *The Guardian*, 2020, https://www.theguardian.com/world/2020/may/10/after-coronavirus-focus-on-the-climate-emergency; Chloé Farand, 'Guterres confront China over coal boom, urging a green recovery', *Climate Home News*, 2020, https://www.climatechangenews.com/2020/07/23/guterres-confronts-china-coal-boom-urging-green-recovery/; and Rupert Read, 'The coronavirus gives humanity one last chance – but for what exactly?', *Compass*, 2020, https://www.compassonline.org.uk/the-coronavirus-gives-humanity-one-last-chance-but-for-what-exactly/.

128 See Read, 'Negotiating the Space Between'.

129 See also Matt McGrath, 'Climate change: 12 years to save the planet? Make that 18 months', *BBC News*, 2019, https://www.bbc.co.uk/news/science-environment-48964736; and Mark Hertsgaard, '"We're losing the race': UN secretary call climate change an "emergency"', *The Guardian*, 2019, https://www.theguardian.com/environment/2019/sep/18/un-secretary-general-climate-crisis-trump.

130 See e.g. Leo Barasi, 'Guest post: Polls reveal surge in concern in UK about climate change', *Carbon Brief*, 2019, https://www.carbonbrief.org/guest-post-rolls-reveal-surge-in-concern-in-uk-about-climate-change for vindication of the claim that XR (on the back of David Attenborough, the Fridays for Future movement, and of course the emerging climate chaos itself) dramatically changed the level of climate-concern in the UK. This concern has not gone away: see e.g. 'Two thirds of Britons believe Climate Change as serious as Coronavirus and majority want Climate prioritised in economic recovery', *Ipsos MORI*, 2020, https://www.ipsos.com/ipsos-mori/en-uk/two-thirds-britons-believe-climate-change-serious-coronavirus-and-majority-want-climate-prioritised.

131 Before you ask: yes, I've done this myself. I've given about ninety thousand pounds to radical green causes from my salary. I've also made the kind of direct asks essayed here before, and had several people respond by giving five-figure sums. I've never done so in writing before, though...

132 I am thinking of the quotation often attributed to Oscar Wilde, and which does indeed seem to fit his 'knowing' wit and aestheticism, which stood in an awkward relation to his avowed socialism: 'The trouble with socialism is that it takes up too many evenings.' ... though, to be fair to Wilde, it is not certain that he actually ever said this remark that is often attributed to him... ('Socialism Would Take Too Many Evenings', *Quote Investigator*, https://quoteinvestigator.com/2019/06/03/evenings/).

133 Rachel Carson, *Silent Spring* (Boston: Houghton Mifflin, 1962).

134 Donella H. Meadows, et. al., *Limits To Growth* (New York: Universe Books, 1972), https://clubofrome.org/publication/the-limits-to-growth/. See Nafeez Ahmed, 'Scientists vindicate "Limits to Growth" – urge investment in "circular economy', *The Guardian*, 2014, https://www.theguardian.com/environment/earth-insight/2014/jun/04/scientists-limits-to-growth-vindicated-investment-transition-circular-economy for the vindication this oft-vilified historic report is now receiving.

135 See Jonathan Watts, 'David Wallace-Wells on climate: "People should be scared – I'm scared"', *The Guardian*, 2019, https://www.theguardian.com/environment/2019/feb/03/david-wallace-wells-on-climate-people-should-be-scared-im-scared.

136 I am borrowing my phrasing from *The Lord of the Rings: The Two Towers* film (2002), when Merry entreats Treebeard not to abandon the forests to the destructivity of Saruman. See 'LOTR The Two Towers

– The Entmoot Decides', *YouTube*, 2014, https://www.youtube.com/watch?v=AXgWZyb_HgE.

137 I make the case that the same is true of the great films that I analyse in my book *A Film-philosophy of Ecology and Enlightenment*; films that include those mentioned earlier: such as *Avatar, The Road, Never Let Me Go* and *Melancholia*. Such films are incomplete without an *active* audience response.

A Proposal

138 Damian Carrington, 'Another two years lost to climate inaction, says Greta Thunberg', *The Guardian*, 2020, https://www.theguardian.com/environment/2020/aug/19/another-two-years-lost-to-climate-inaction-says-greta-thunberg.

139 In her impassioned, enraged speech at the UN in September 2019: Bill Chappell, '"This Is All Wrong," Greta Thunberg Tells World Leaders at U.N. Climate Session', *NPR*, 2019, https://www.npr.org/2019/09/23/763389015/this-is-all-wrong-greta-thunberg-tells-world-leaders-at-u-n-climate-session. The key passage runs, *'…this is all wrong. I shouldn't be up here. I should be back in school, on the other side of the ocean. Yet you all come to us young people for hope. How dare you?* You have stolen my dreams, and my childhood, with your empty words. And yet I'm one of the lucky ones … People are suffering. People are dying. Entire ecosystems are collapsing. We are in the beginning of a mass extinction, and all you can talk about is money, and fairy tales of eternal economic growth. How dare you?!" (emphasis added).

140 And if striking every Monday didn't work, then perhaps the strikes would be expended to Tuesdays as well; and… you see where this would head, and how as a strategy it would bring vast pressure on governments and employers.

141 It will probably once more occur to readers that Parents for a Future, while massively inclusive, will not straightforwardly include the childless. My response to this valid concern is implicit in the three-fold typology of the childless I gave already in Chapter 2. In very brief: those who are childless *so as to* devote themselves to the cause of the future will support PFAF anyway; those who love their nieces and nephews etc, likewise; while by contrast those who truly don't care about the long future can and should be ignored.

I myself fall into both the first two categories. Perhaps there is a case for some small 'satellite' organisations; perhaps I myself might head up 'Aunts and Uncles for a future'; perhaps others might create 'Foster-parents for a future' or 'Guardians for a future'. These would all fit under the broader umbrella of 'Humans [or 'People'] for a future'. Which would connote the point, of everyone being called to work together for a future for our descendants. But it would *miss* the thrust of this book to think that that umbrella is what counts. No, *parenting* the future is the heart of the matter; and that's mostly based in what most of us adult humans are: parents. It is in the as-yet largely untapped appeal to parents that the novel and consequential thrust of my essay is most strongly found. We non-parents should strongly support parents; but if parents were *en masse* to really start taking the existential crisis facing the future seriously, they would move mountains.

142 I should point out that I learnt in the very final stages of writing this book of the existence of an excellent growing international organisation called Parents For Future. This began as a support-group for Fridays For Future but is on the way to becoming much more than that. I look forward to working with PFF and others in the burgeoning space of parental concern re the eco-emergency (including the excellently named nascent 'Mothership') on the ideas contained in this Proposal...

143 See my '24 Theses on Corona' for analysis.

144 See Rupert Read and Dario Kenner, 'XR UK: Telling the truth through targeted disruption', *openDemocracy*, 2019, https://www.opendemocracy.net/en/opendemocracyuk/xr-uk-telling-truth-through-targeted-disruption/; and Rupert Read, 'How a movement of movements can win: Taking XR to the next level', *rupertread.net*, 2019, https://rupertread.net/writings/2019/how-movement-movements-can-win-taking-xr-next-level for some further explication of this crucial point about how a movement could be broad-based, universalistic, as a true emergency-response.

145 See Read, 'An introduction to transformative adaptation' for what I mean by this term. Which is sometimes bolder than the 'mainstream' understanding of the term; for some discussion, see Roger Few, et. al., 'Transformation, adaptation and development: relating concepts to practice', *Palgrave Communications* no. 3, 2017, https://www.nature.com/articles/palcomms201792.

146 This is if anything putting it mildly. For a view more pessimistic than mine on how very too late we have left it, see this letter co-authored by major climate scientists: 'After coronavirus, focus on the climate emergency: Letters', *The Guardian*, 2020, https://www.theguardian.com/world/2020/may/10/after-coronavirus-focus-on-the-climate-emergency. See also Hans Joachim Schellnhuber, 'Foreword', in *What Lies Beneath: The Understatement of Existential Climate Risk*. And this study by Prof. Kevin Anderson, et. al.: 'UK & Sweden's carbon targets half what is needed'; the scale of emissions cuts needed in 'developed' (sic) countries is far outside the realms of politics as usual, exceeding even what happened in 2020, the year of the 'Covid pause'.

147 See Chapter 3, 'Making the Best of Climate Disasters', in *Facing up to Climate Reality.*

Acknowledgements

I owe deep thanks to Ruth Makoff for thinking the topic of Chapter 3 with me. To Nassim N. Taleb for co-thinking matters precautionary with me across the years (thus the whole book, but especially Chapter 4). To Adam Woodhall, Jem Bendell and Joel Scott-Halkes for inspiration that helped bring together my Proposal. To Philip Langeskov for being a brilliant editorial presence and Samantha Purvis for being a brilliant editorial assistant. To Tara Greaves for help throughout the manuscript. To Vlad Vexler, Gudrun Freese and Marcus Hemsley for key help in framing. To Peter Kramer, Victor Anderson, John Foster, Kate Rawles and Tim O'Riordan for readings of drafts of the manuscript. To UEA for repeated funding, time and support that has helped make this book possible and make it as good as it is (if it is).

Thanks also to Skeena Rathor, Jonathan Kent, Molly Scott Cato, Josie Collins, Atus Mariqueo-Russell, Jasmin Kirkbride, Andrew Boswell, Roc Sandford, Adrian Ramsay, Caroline Lucas, Jonathon Porritt, Samantha Earle, Tom Greaves, Jamie Kelsey-Fry, Rebecca Gibbs, Deepak Rughani, Emily Grossman, Jane Morton, Alistair MacIntosh, Lucy Hinton, Jeremy Thres, Charles Eisenstein, Ronan Harrington, Liam Kavanagh, Svante Thunberg, Nathan Hamilton, Martha Griffiths, Senica Maltese, James Andow, and too many colleagues in XR and in UEA to mention. Thank you also to Wilfrid, who is seven, and Arthur, who is five, for their help with the cover designs. We hope they enjoy their LEGO. Finally, obviously, to Juliette Harkin, with love; and to my nieces, Poppy and Rosie, and my nephew Joe, for keeping my eyes focused on the prize, on nurturing the future.

The Author

Professor Rupert Read is based in the Philosophy Department at the University of East Anglia. He is widely known for having got the BBC to change its policy on the reporting of dangerous human-caused climate-change (in 2018), such that the BBC no longer features climate-deniers to 'balance' the facts. He has been a national spokesperson for Extinction Rebellion (appearing on *Today*, *Question Time*, and many more), and for the Green Party, and was formerly a two-term elected Green Party local Councillor in Norwich. He co-convened Extinction Rebellion's Political Liaison Group, meeting on several occasions with the UK Government in 2019. He is an expert on the Precautionary Principle, on which he has won AHRC grants and written reports for Parliamentarians. He is author of *Philosophy for Life: Applying Philosophy in Politics and Culture* (2007), *This Civilisation is Finished: Conversations on the end of Empire and What Lies Beyond* (2019) and *Extinction Rebellion: Insights from the Inside* (2020), and co-author with colleagues in Green House, the thinktank he co-founded, of *Facing up to climate reality: Honesty, Disaster and Hope* (2019). His next book will be *Why Climate Breakdown Matters*, with Bloomsbury Press.

Parents For A Future
How Loving Our Children Can Prevent Climate Collapse
By Rupert Read

First published in this edition
UEA Publishing Project Ltd., 2021
Copyright © Rupert Read, 2021

The right of Rupert Read to be identified as the Author of
this work has been asserted by him in accordance with the
Copyright, Design & Patents Act, 1988.

Design and typesetting by Louise Aspinall
Typeset in Arnhem Pro
Printed by Imprint Digital
Distributed by NBN International
This book is printed on recycled paper

ISBN 978-1-911343-37-0

This is just the beginning... To help this movement to expand democracy and the rights of future generations, please visit www.parentsforafuture.org and sign up to alerts.

@Parents4aFuture #ParentsForaFuture